# FEDERAL LAND TRANSACTION FACILITATION ACT

# FEDERAL LAND TRANSACTION FACILITATION ACT

## GOVERNMENT ACCOUNTABILITY OFFICE

**Novinka Books**
*New York*

For permission to use material from this book please contact us:
Telephone 631-231-7269; Fax 631-231-8175
Web Site: http://www.novapublishers.com

### NOTICE TO THE READER

The Publisher has taken reasonable care in the preparation of this book, but makes no expressed or implied warranty of any kind and assumes no responsibility for any errors or omissions. No liability is assumed for incidental or consequential damages in connection with or arising out of information contained in this book. The Publisher shall not be liable for any special, consequential, or exemplary damages resulting, in whole or in part, from the readers' use of, or reliance upon, this material.

This publication is designed to provide accurate and authoritative information with regard to the subject matter covered herein. It is sold with the clear understanding that the Publisher is not engaged in rendering legal or any other professional services. If legal or any other expert assistance is required, the services of a competent person should be sought. FROM A DECLARATION OF PARTICIPANTS JOINTLY ADOPTED BY A COMMITTEE OF THE AMERICAN BAR ASSOCIATION AND A COMMITTEE OF PUBLISHERS.

LIBRARY OF CONGRESS CATALOGING-IN-PUBLICATION DATA

Available Upon Request
ISBN 978-1-60692-056-5

*Published by Nova Science Publishers, Inc.* ≃ *New York*

# CONTENTS

| | | |
|---|---|---|
| **Correspondence** | | vii |
| **Abbreviations** | | xi |
| **Chapter 1** | Results in Brief | 1 |
| **Chapter 2** | Background | 5 |
| **Chapter 3** | BLM Has Raised Most FLTFA Revenue from Land Sales in Nevada | 11 |
| **Chapter 4** | BLM Faces Several Challenges to Future Sales under FLTFA | 17 |
| **Chapter 5** | Agencies Have Purchased Few Parcels with FLTFA Revenue | 27 |
| **Chapter 6** | Agencies Face Challenges in Completing Additional Acquisitions | 43 |
| **Chapter 7** | Conclusions | 51 |
| **Chapter 8** | Matters for Congressional Consideration | 53 |
| **Appendix I** | Objectives, Scope, and Methodology | 55 |
| **Appendix II** | Completed FLTFA Land Sales, through May 2007 | 59 |

**Appendix III**      Detailed Information on Planned
                     FLTFA Land Sales through 2010,
                     as Reported by BLM State Offices                77

**References**                                                       85

**Index**                                                           89

# CORRESPONDENCE

February 5, 2008

The Honorable Norman D. Dicks
Chairman
The Honorable Todd Tiahrt
Ranking Member
Subcommittee on Interior, Environment, and Related Agencies
Committee on Appropriations
House of Representatives

The four major federal land management agencies—the U.S. Department of the Interior's Bureau of Land Management (BLM), Fish and Wildlife Service, and National Park Service, and the U.S. Department of Agriculture's Forest Service—administer approximately 628 million acres, or about 28 percent of the land area in the United States. These public lands are mostly in the 11 western states and Alaska, where the four agencies manage lands that constitute significant portions of the states' acreage, ranging from about 28 percent in Washington state to about 81 percent in Nevada. These lands have multiple uses, from preserving cultural and natural treasures to accommodating the development of resources, such as oil and gas, among other things. Historically, many controversies have arisen over the agencies' management of these lands, including the selling of federal land and the purchasing of private land. In these controversies, the agencies have had to balance the need to protect resources in the land they manage with the need to respect the rights of private landowners.

One particularly controversial issue has been managing federal lands with inholdings, which are nonfederal lands within the boundaries of national parks, forests, wildlife refuges, and other designated areas. In 2005, the agencies estimated there were at least 70 million acres of inholdings within the lands they manage [1]. Inholdings can create significant management problems for federal agencies in maintaining boundaries, providing security, and protecting resources, among other things. The federal land management agencies have had the authority to acquire inholdings, but have had limited funding for exercising this authority.

Congress enacted the Federal Land Transaction Facilitation Act of 2000 (FLTFA), in part, to enhance the efficiency and effectiveness of federal land management by allowing the four land management agencies to acquire inholdings to promote the consolidation of ownership of public and private lands in a manner that would allow for better overall resource management [2]. Revenue generated by the sale or exchange of public lands under FLTFA has created another funding source available to the four agencies to acquire land when appropriations for acquisitions have been declining [3]. These funds are available to the agencies without further appropriation.

BLM, which manages approximately 256 million acres of federal land, is authorized to sell or exchange land identified in its land use plans [4, 5]; the other three land management agencies have limited or no sales authority. Therefore, the funds for FLTFA acquisitions must come from the revenue generated by BLM sales or equalization payments derived from exchanges. BLM may dispose of land that meets certain criteria, including land that is difficult to manage, no longer needed, or needed for community expansion. Thus, when BLM sells land, the sale generates revenue and reduces the burden on its land managers to accomplish such tasks as monitoring scattered acreage and boundaries.

Once BLM has sold land, FLTFA directs BLM to deposit the revenue generated from these transactions into a special U.S. Treasury account created by FLTFA [6]. However, the act limits the revenue deposited into this account to that generated from sales or exchanges of public lands identified for disposal in a land use plan in effect as of July 25, 2000—the date of FLTFA's enactment. Money in the new account is available to BLM and the other three agencies to purchase inholdings, and in some cases, land adjacent to federally designated areas and containing exceptional resources [7]. The act expires in July 2010, and the Administration has proposed revising and extending it.

BLM sells its land in one of three ways: competitive sales; modified competitive sales, which provide a preference to existing land users or adjoining landowners; and direct sales, which occur in special situations, such as when parcels are completely surrounded by one landowner and there is no public access. BLM prefers competitive sales because these usually generate the most revenue and, therefore, are more likely to increase the revenue available under FLTFA for land acquisitions. BLM staff in headquarters, its 12 state offices, and 144 field offices nationwide manage and conduct these sales [8]. About 300 full-time equivalent staff, out of a workforce of about 10,500 full-time equivalent staff, are responsible for land and realty management in BLM. These staff are directly responsible for land sales and acquisitions, along with other realty responsibilities, such as processing energy rights-of-way and leasing and permitting on public lands.

The federal land agencies have two methods for identifying land to acquire under FLTFA. First, the agencies can nominate parcels through a process laid out in state-level implementation agreements that were developed under the direction of a national memorandum of understanding (MOU) that implemented the program. Under the process, state-level interagency teams are to review proposals for land acquisitions and forward their nominations to the Secretaries of Agriculture and of the Interior for approval. Second, the Secretaries can directly use a portion of FLTFA revenue to acquire specific parcels of land at their own discretion. The national MOU laid out the expectation that most acquisitions would occur through the state-level process.

FLTFA places several restrictions on using funds from the new U.S. Treasury account. Among other things, FLTFA requires that (1) no more than 20 percent can be used for BLM's administrative and other activities necessary to carry out the land disposal program; (2) of the amount not spent on administrative expenses, at least 80 percent of the revenue must be expended in the state in which the funds were generated; and (3) at least 80 percent of FLTFA revenue required to be spent on land acquisitions within a state must be used to acquire inholdings (as opposed to adjacent land) within that state. In addition, the national MOU sets the allocation of funds from the FLTFA account for each agency—60 percent for BLM, 20 percent for the Forest Service, and 10 percent each for the Fish and Wildlife Service and the Park Service, but the Secretaries may vary from these allocations by mutual agreement.

With FLTFA expiring in July 2010, you asked us to (1) determine the extent to which BLM has generated revenue for the FLTFA program, (2) identify challenges BLM faces in conducting future sales, (3) determine the extent to which agencies have spent funds under FLTFA, and (4) identify challenges the agencies face in conducting future acquisitions.

To address these objectives, we reviewed FLTFA, other applicable authorities, and agency guidance, and interviewed FLTFA program leads at the four agencies' headquarters, officials with Interior's Office of the Solicitor, and officials with the Office of the Assistant Secretary for Land and Minerals Management on FLTFA implementation. We also obtained and analyzed data from BLM's Division of Business Services on program revenue and expenditures and visited the Division of Business Services accounting officials in Lakewood, Colorado, to discuss the management of the FLTFA account [9]. We conducted semistructured interviews and collected data from (1) the 10 BLM state officials responsible for the FLTFA program in their office on the program's status, completed and planned FLTFA land sales and acquisitions, and the challenges faced in conducting sales and acquiring land; (2) officials at the 7 BLM field offices that have raised 97 percent of the FLTFA revenue; and (3) a nongeneralizable sample of 11 of the 137 remaining BLM district and field offices that had not conducted a competitive sale under FLTFA as of May 31, 2007 to determine why such sales have generally not occurred and challenges faced to conducting future sales. With regard to acquisitions, we reviewed available documentation for land acquisition proposals considered by the 10 FLTFA interagency teams at the state level, agency headquarters, and the Secretaries of Agriculture and of the Interior. During our visits to BLM state offices (California, Nevada, New Mexico, and Oregon) and field offices (Carson City, Nevada, and Las Cruces, New Mexico), we interviewed officials and visited planned land acquisition sites to learn about the details of the land acquisition process. A more detailed description of our scope and methodology is presented in appendix I. We performed our work between November 2006 and February 2008 in accordance with generally accepted government auditing standards. Those standards require that we plan and perform the audit to obtain sufficient, appropriate evidence to provide a reasonable basis for our findings and conclusions based on our audit objectives. We believe that the evidence obtained provides a reasonable basis for our findings and conclusions based on our audit objectives.

# ABBREVIATIONS

| | |
|---|---|
| BLM | Bureau of Land Management |
| FLPMA | Federal Land Policy and Management Act of 1976 |
| FLTFA | Federal Land Transaction Facilitation Act of 2000 |
| LWCF | Land and Water Conservation Fund |
| MOU | memorandum of understanding |
| SNPLMA | Southern Nevada Public Land Management Act of 1998 |

*Chapter 1*

# RESULTS IN BRIEF[*]

Since FLTFA was enacted in 2000, BLM has raised $95.7 million in revenue, mostly from selling 16,659 acres. As of May 2007, about 92 percent of the revenue raised, or $88 million, has come from land sales in Nevada. Revenue grew slowly during the first years of the program and peaked in fiscal year 2006, when a total of $71.1 million was generated. BLM's Nevada office accounts for the lion's share of the sales because (1) demand for land to develop has been high in rapidly expanding population centers such as Las Vegas, (2) BLM has a high percentage land in proximity to these centers, and (3) BLM has experience selling land under another federal land sales program authorized for southern Nevada. More specifically, the Carson City and Las Vegas Field Offices generated a total of $86.2 million, or 90 percent of all revenue generated under FLTFA, mostly through a few competitive sales. As of May 31, 2007, BLM offices covering three other states—New Mexico, Oregon, and Washington—have raised over $1 million each, and the remaining seven BLM state offices—Arizona, California, Colorado, Idaho, Montana, Utah, and Wyoming—had each raised less than $1 million. Most BLM field offices have not generated revenue under FLTFA.

BLM faces several challenges to raising revenue through future sales under FLTFA that BLM managers and we identified. Most frequently, BLM state and field officials cited the lack of availability of knowledgeable realty staff to conduct the sales as a challenge. These staff may not be available because they are working on activities that BLM has identified as higher priorities, such as reviewing and approving energy rights-of-way. This

---

[*] This book is an excerpted indexed version of GAO Report GAO-08-196, Dated February 2008

challenge is followed, in the order of frequency cited, by the time, cost, and complexity of the land sales process; external factors, such as public opposition to a sale; FLTFA program and legal restrictions; and the land use planning process. Many of the challenges they raised to conducting sales are not unique to FLTFA sales. We identified two additional issues hampering land sales activity under FLTFA. First, although BLM has identified land for sale in its land use plans, it has not made the sale of these lands a priority during the first 7 years of the program. Furthermore, BLM has not set goals for FLTFA sales or developed a sales implementation strategy. The establishment of goals is an effective management tool for measuring and achieving results. While some BLM state offices told us they have planned FLTFA sales—96, totaling 25,404 acres—through 2010, BLM has no overall implementation strategy for generating funds to purchase inholdings as mandated by FLTFA. Second, although BLM has identified a number of land parcels for disposal since the act's passage, revenue from these potential sales will not be eligible for deposit into the FLTFA account because the act only allows the deposit of revenue from the sale of land identified for disposal on or before July 25, 2000, the date of its enactment.

BLM reports that the four land management agencies have spent $13.3 million of the $95.7 million in the FLTFA account. They spent $10.1 million to acquire nine parcels totaling 3,381 acres in seven states—Arizona, California, Idaho, Montana, New Mexico, Oregon, and Wyoming. In addition, BLM spent $3.2 million for administrative expenses between 2000 and 2007 to conduct FLTFA-eligible sales, primarily in Nevada. The agencies acquired the land between August 2007 and January 2008—more than 7 years after FLTFA was enacted. These acquisitions were initiated using the Secretaries' discretion, and most had been identified but not funded for purchase under another land acquisition program. As of October 2007, no land had been purchased through the state-level interagency nomination process that was established by the national MOU and state agreements. The agencies envisioned these agreements as the primary process for acquiring land under FLTFA. Acquisitions have not yet occurred under the state-level process because it has taken 6 years to complete the interagency agreements needed to implement the program and because relatively little revenue is available for acquisitions outside of Nevada, owing to the FLTFA requirement that, excluding administrative expenses, at least 80 percent of the funds must be spent in the state where revenue were raised. Although Nevada has proposed five acquisitions, none have been completed. Two of the proposed acquisitions approved by the secretaries failed because of differences with the owners, one was withdrawn because it did not meet FLTFA criteria, one is pending secretarial approval, and one was recently approved.

BLM managers and we identified several challenges to completing future land acquisitions under FLTFA. Most frequently, BLM state and field officials cited the time, cost, and complexity of the acquisition process as a challenge. For example, to complete an acquisition under the MOU, four agencies must work together to identify, nominate, and rank proposed acquisitions, which must then be approved by the two Secretaries. The other most commonly cited challenges officials raised were, in order of frequency, (1) identifying a willing seller, (2) the availability of knowledgeable staff to conduct acquisitions, (3) the lack of funding to purchase land, (4) restrictions imposed by laws and regulations, and (5) public opposition to land acquisitions. Some of these challenges are likely typical of many federal land acquisitions. Officials from the other three agencies had few comments on challenges to acquisitions because they have had little experience with the program. We also found that the act's restriction on the use of funds outside of the state in which they were raised continues to limit acquisitions. Specifically, little revenue is available for acquisitions outside of Nevada. Furthermore, progress in acquiring priority land has been hampered by the agencies' weak performance in identifying inholdings and setting priorities for acquiring them, as required by the act. In addition, we found that the agencies have not established procedures to track key provisions in the act and the national MOU. Specifically, the agencies have not established a procedure to track the act's requirement that at least 80 percent of FLTFA revenue allocated for land acquisitions in each state are used to acquire inholdings in that state. In addition, BLM has not established a procedure to track agreed-upon fund allocations—60 percent for BLM, 20 percent for the Forest Service, and 10 percent each for the Fish and Wildlife Service and the Park Service in the national MOU. Because the agencies have not tracked these amounts, they cannot ensure they are fully complying with the act or fully implementing the MOU.

If Congress decides to reauthorize FLTFA, we raise two matters for congressional consideration to better meet the goals of FLTFA. These matters relate to making additional land eligible for sales and increased flexibility in the use of funds for acquisitions under the program. In addition, we are making five recommendations to the Secretaries to better manage and oversee the FLTFA program, such as developing goals for FLTFA sales and a strategy to implement them. In commenting on a draft of this book the Department of the Interior generally concurred with our findings and recommendations, stating that it will implement all of the recommendations.

Their comments are presented in appendix IV of this book. In addition, Interior and the Department of Agriculture provided technical comments on the draft book, which we have incorporated as appropriate.

# Chapter 2

# BACKGROUND

FLTFA, commonly called the "Baca Act," [10] provides for the use of revenue from the sale or exchange of BLM land identified for disposal under land use plans in effect as of the date of its enactment—July 25, 2000. The act does not apply to land identified for disposal after its enactment, such as through a land use plan amendment approved after that date. Revenue generated under FLTFA are available to the Secretaries of Agriculture and of the Interior for acquiring inholdings within certain federally designated areas, or land adjacent to those areas and containing exceptional resources, and for administrative and other expenses necessary to carry out the land disposal program under the FLTFA.

To implement FLTFA, BLM has designated a program lead realty specialist in headquarters, in each state office involved, and in each field office within those states. The program lead duties are sometimes split between land and realty staff who specialize in sales and others who specialize in the acquisition process. In addition, to facilitate the use of FLTFA funds for acquisition, the other three agencies sharing in the revenue, the Forest Service, the Park Service, and the Fish and Wildlife Service, have also designated realty staff to participate in interagency groups to decide on acquisitions in each BLM state. BLM manages the FLTFA account through its Division of Business Services.

# FEDERAL LAND SALES AUTHORITIES AND PROCESS

Although FLTFA authorizes proceeds from eligible land sales and exchanges to be used in acquiring land, it does not provide any new sales authority. The sales authority, as stated in FLTFA, is provided by the Federal Land Policy and Management Act of 1976 (FLPMA). FLPMA authorizes the Secretary of the Interior to dispose of certain federal lands—through sale and exchange, among other disposal methods—and authorizes the Secretaries of Agriculture and of the Interior to acquire certain nonfederal lands. FLPMA also authorizes the Secretary of Agriculture to exchange land. FLPMA requires the Secretary of the Interior to develop land use plans to determine which lands are eligible for disposal and acquisition. The level of specificity differs in land use plans, from describing general areas to naming specific parcels. In developing these land use plans, agencies must work closely with federal, state, and local governments and allow for public participation. Land use plans are typically revised every 15 to 20 years to address changing land use conditions in the area covered.

Sales and acquisitions must comply with requirements of FLPMA and other applicable laws, which can require, among other things, an assessment of the environmental impacts of the proposed land transaction, assessment of natural and cultural resources, preparation of appraisals, and public involvement. Furthermore, with regard to land sales specifically, FLPMA requires that land be sold at the appraised fair market value or higher.

Although BLM policy states that competitive sales are preferred when a number of parties are interested in bidding on a parcel for sale, regulations for the FLPMA land sales authority provide for other methods of sale when certain criteria are met. The regulations state that modified competitive sales may be used to permit the current grazing user or adjoining landowner to meet the high bid at the public sale. This procedure allows for limited competitive sales to protect ongoing uses, to assure compatibility of the possible uses with adjacent land, and to avoid dislocating current users. The regulations state that a direct sale may be used when the land offered for sale is completely surrounded by land in one ownership with no public access, when the land is needed by state or local governments or nonprofit corporations, or when the land is necessary to protect current equities in the land or resolve inadvertent unauthorized use or occupancy of the land.

In completing the steps necessary to purchase land, third-party organizations, such as The Nature Conservancy and The Trust for Public Land, often provide assistance to the federal government. For example, third parties may assist by purchasing desired land for eventual resale to the federal government or by negotiating an option with the seller to purchase land within a specified period of time, which provides additional time for the federal agency to secure necessary funding for the purchase or to comply with laws and regulations governing the acquisition.

## FEDERAL LAND ACQUISITION FUNDING

The primary source for land acquisition funding for BLM, the Park Service, the Forest Service, and the Fish and Wildlife Service, has traditionally been the Land and Water Conservation Fund (LWCF), which was created to help preserve, develop, and assure access to outdoor recreation resources [11]. To receive LWCF funding, the agencies independently identify and set priorities for land acquisitions and then submit their list of priority acquisitions in their annual budget request to Congress. LWCF funding is available for land acquisition purposes only if appropriated by Congress, unlike the funds in the FLTFA account, which are available without further appropriation.

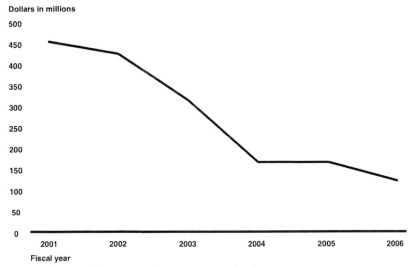

Source: GAO analysis of Congressional Research Service data.

Figure 1. LWCF Land Acquisition Appropriations, Fiscal Year 2001 through Fiscal Year 2006.

LWCF land acquisition appropriations have been declining in recent years. Specifically, funds for the four agencies declined from $453.4 million appropriated in fiscal year 2001 to $120.1 million appropriated in fiscal year 2006, as depicted in figure 1 [12]. BLM has traditionally received the lowest amount of LWCF land acquisition funding among the four agencies. For example, in fiscal year 2006, BLM's share of total appropriated LWCF land acquisition funding was only $8.6 million, or about 7 percent of the total appropriation [13]. BLM's land sales eligible under FLTFA have created another funding source for the four agencies to acquire land. FLTFA provides that if all funds in the account are not used by the sunset date in 2010, they will become available for appropriation under section 3 of the Land and Water Conservation Fund Act.

## OTHER LAND SALE LAWS

Other laws allow BLM to retain certain proceeds from federal land sales and share them among agencies for land acquisitions, as well as other purposes. The most notable of these is the Southern Nevada Public Land Management Act of 1998 (SNPLMA) [14]. SNPLMA's stated purpose is to "provide for the orderly disposal of certain federal lands in Clark County, Nevada, and to provide for the acquisition of environmentally sensitive land in the State of Nevada" [15]. Since enactment, SNPLMA has generated just under $3 billion in revenue. As of September 2007, a portion of this revenue has been spent, in part, to complete 41 land acquisition projects in Nevada for a total of $129.1 million. Unlike FLTFA, SNPLMA has no expiration date and its sales receipts are placed in an interest bearing account. However, it has fewer acres available for disposal than FLTFA.

## FLTFA REQUIREMENTS ON USE OF REVENUE AND OTHER KEY PROVISIONS

FLTFA places a number of requirements on the use of revenue generated under the act. Among these requirements, BLM must provide 4 percent of sale proceeds to the state in which revenue was raised for education and transportation purposes [16]. Figure 2 illustrates these requirements using an example of $1 million in revenue.

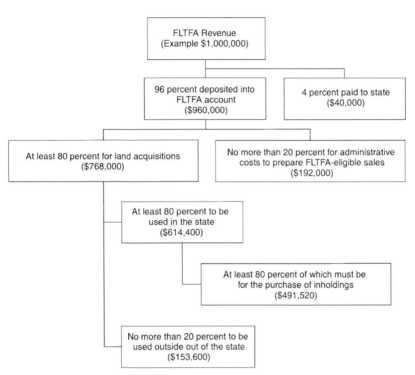

Source: GAO analysis of FLTFA and BLM's Instruction Memorandum No. 2007-205,
September 25, 2007.

Note: With the exception of the amount designated for inholdings, the steps depicted in this
figure under the 96 percent deposited into the FLTFA account are subactivity accounts
established in BLM's Federal Financial System to record the fund allocations as
authorized by FLTFA.

Figure 2. Requirements for Using FLTFA Revenue.

FLTFA also limits land acquisitions to land within and adjacent to
federally designated areas, such as national parks, national forests, and
national conservation areas. While most lands managed by the Fish and
Wildlife Service, the Forest Service, and the Park Service are federally
designated areas, many of the lands managed by BLM are not federally
designated areas; therefore, acquisitions within undesignated lands would
not qualify under FLTFA.

Furthermore, FLTFA requires that the Secretaries establish a procedure
to identify and set priorities for acquiring inholdings. As part of this process,
it called for the Secretaries to consider (1) the date the inholding was
established, (2) the extent to which the acquisition would facilitate

management efficiency, and (3) other criteria identified by the Secretaries. The act also requires a public notice be published in the *Federal Register* detailing the procedures for identifying inholdings and setting priorities for them and other information about the program.

# MEMORANDUM OF UNDERSTANDING IMPLEMENTS FLTFA

To improve FLTFA implementation, the four agencies signed a national MOU. Among other things, the MOU established a Land Transaction Facilitation Council, which consists of the heads of the four agencies and the U.S. Department of the Interior Assistant Secretary for Policy, Management, and Budget to oversee the implementation and coordination of activities undertaken pursuant to the MOU. The MOU also directed the agencies to establish state-level implementation plans that would establish roles and responsibilities, procedures for interagency coordination, and field-level processes for identifying land acquisition recommendations and setting priorities for these recommendations.

# PROPOSED AMENDMENTS TO FLTFA

The Administration has proposed revising and extending the act.

Specifically, the U.S. Department of the Interior's fiscal year 2007 and 2008 budgets included proposals to

- allow BLM to use updated land use plans to identify areas suitable for disposal,
- allow a portion of receipts to be used by BLM for restoration projects,
- require BLM to return 70 percent of net proceeds from eligible sales to the U.S. Treasury, and
- cap retention of receipts at $60 million per year.

In addition, the U.S. Department of the Interior called for Congress to extend the FLTFA program to 2018.

# Chapter 3

# BLM HAS RAISED MOST FLTFA REVENUE FROM LAND SALES IN NEVADA

Since FLTFA was enacted in 2000, BLM has raised $95.7 million in revenue, mostly by selling 16,659 acres. As of May 2007, about 92 percent of the revenue raised, or $88 million, has come from land sales in Nevada— 1 of the 11 western states under FLTFA. Nevada accounts for most of the sales because of rapidly expanding population centers coupled with a high percentage of BLM land in the state and experience selling land under the SNPLMA program. Most BLM field offices have not generated revenue under FLTFA.

## BLM HAS RAISED $95.7 MILLION FROM FLTFA LAND SALES, PRIMARILY IN TWO NEVADA FIELD OFFICES

Between July 2000 and May 2007, BLM raised $95.7 million in revenue for selling 16,659 acres, according to data verified by BLM state offices. In addition, the BLM Division of Business Services reports exchange equalization payments totaling $3.4 million. Nevada has accounted for the great majority of the sales. As of May 2007, about 92 percent of the revenue raised, or $88 million, has come from land transactions in Nevada. More specifically, the Carson City and Las Vegas field offices generated a total of $86.2 million, or 90 percent of all revenue generated under FLTFA, mostly through a few competitive sales. For example, the Carson City Field Office raised $39.1 million through 3 sales and Las Vegas Field Office raised $33.6 million through 7 sales. Table 1 shows the state-by-state totals of sales revenue generated, acres sold, and number of sales. See appendix II for a listing of completed sales BLM state offices have reported to us.

**Table 1. BLM Reported FLTFA Cumulative Revenue from Sales by State, July 25, 2000, through May 31, 2007**

| State | Cumulative sales revenue | Acres sold | Number of sales |
|---|---|---|---|
| Arizona | 54,102 | 28 | 2 |
| California | 235,010 | 215 | 10 |
| Colorado | 939,856 | 243 | 25 |
| Idaho | 180,740 | 206 | 9 |
| Montana | 59,000 | 53 | 3 |
| Nevada | 88,010,041 | 5,399 | 106 |
| New Mexico | 4,052,800 | 778 | 14 |
| Oregon (includes Washington state) | 1,103,485 | 8,501 | 82 |
| Utah | 177,000 | 26 | 1 |
| Wyoming | 840,085 | 1,209 | 13 |
| **Total** | **$95,652,119** | **16,659** | **265** |

Source: GAO analysis of BLM Division of Business Services data verified by BLM state offices.

Notes: Numbers may not add due to rounding. The revenue numbers provided in this table are before the 4 percent payment to states has been deducted and include only land sale revenue. The Division of Business Services reports exchange equalization payments in these states totaling about $3.4 million, which are also available for FLTFA land acquisitions.

Some of BLM's Nevada field offices, particularly Las Vegas and Carson City, have been in a unique position to raise the most funds under FLTFA to date because of rapidly expanding populations, development in those areas, and the availability of nearby BLM land. In addition, BLM Nevada staff had previous experience with SNPLMA, the land sales program in the Las Vegas area. In fact, the Nevada office used procedures and staff from this program to initiate FLTFA's sales and acquisition programs. According to Nevada state office officials, BLM's annual work plan for lands and realty work specifically directed the Nevada office to continue to hold FLTFA and SNPLMA land sales as appropriate.

Revenue from land sales and exchanges under FLTFA grew slowly in the first years of the program but picked up in fiscal years 2004 and 2005, with $16.6 million and $4.8 million, respectively. Revenue reached a peak in fiscal year 2006, when a total of $71.1 million was collected. BLM officials said the land sales market in Nevada has cooled since its peak in 2006. Figure 3 shows the FLTFA revenue through May 2007.

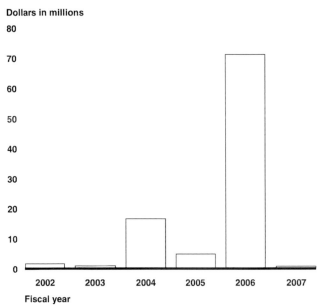

**Dollars in millions**

**Fiscal year**

Source: GAO analysis of BLM's Division of Business Services data verified by BLM state offices.

Notes: Fiscal year 2007 reflects revenue collected through May 31. Since buyers have 180 days to make final payment, some collections are for sales that occurred in the prior fiscal year. Because the FLTFA account was not established until February 2002, funds collected from sales and exchanges in FY 2000 and FY 2001 are included in FY 2002. In addition, we aggregated the revenue for each sale as of the most recent collection date (e.g., if a particular land sale had collections in 2003 and 2004, the total amount collected was included in the 2004 total).

Figure 3. FLTFA Revenue by Fiscal Year, through May 31, 2007

The FLTFA account benefits from the proceeds of all types of transactions, including land exchanges and land sales made on a competitive, modified competitive, or direct basis. BLM sets the appraised fair market value as the sales price for direct sales and as the minimum bid price for competitive sales. Of the 265 completed sales reported by BLM state offices, 149 were competitive, 33 were modified competitive, and 83 were direct. In terms of FLTFA revenue, the great majority, about 96 percent, has been raised from competitive sales. For example, in December 2005, the Las

Vegas Field Office sold a 40-acre parcel through a competitive auction for $7.3 million, or 152 percent of its appraised fair market value of $4.8 million. On a much smaller scale in a December 2006 competitive auction, the Burns District Office in Oregon sold 240 acres for $47,000, or 163 percent of its appraised fair market value of $28,800. In 2006, the Carson City Field Office offered two parcels of about 100 and 106 acres with appraised fair market values of $10 million and $6.4 million, respectively, in north Douglas County, Nevada, just south of the Carson City limits. The former BLM parcels are contiguous and across a major highway from three shopping centers. Through competitive auctions, BLM received final prices of $16.1 million and $8.4 million, or 161 and 131 percent, respectively, of appraised value. Figure 4 shows areas in these two parcels.

Source: BLM Carson City Field Office.
Notes: In the 106-acre parcel on the left, the area left of the roadway was zoned for commercial development. In the 100-acre parcel on the right, the land to the right of the roadway was zoned primarily for residential development.

Figure 4. Two BLM Parcels Near Carson City, Nevada, Sold for a Total of $24.5 Million.

According to a GAO analysis of data from BLM's Division of Business Services and BLM state offices on land sales revenue collected in the FLTFA account, only 12 of 144 field offices have conducted competitive sales. An additional 28 field offices have generated FLTFA revenue through direct or modified competitive sales. The remaining 104 offices have not generated sales revenue for the FLTFA account. Table 2 shows FLTFA sales by the method used and the amount of revenue generated.

Using the data provided by BLM state offices on completed FLTFA sales as of May 31, 2007, we determined that the actual sales prices of the parcels sold exceeded the appraised fair market value of those parcels by 52 percent.

**Table 2. Number of Sales and Revenue Raised by Competitive, Modified Competitive, and Direct Sales under FLTFA, as of May 31, 2007**

Dollars in millions

| Method | Number of sales | Revenue collected | Number of field offices reporting |
|---|---|---|---|
| Competitive | 149 | $91.4 | 12 |
| Modified competitive | 33 | 1.0 | 6 |
| Direct | 83 | 3.2 | 33 |
| **Total** | **265** | **$95.7** | a |

Source: GAO analysis of BLM's Division of Business Services data verified by BLM state offices.

Note: Revenue column does not add due to rounding.

aNo total reported because each field office may use more than one method of sale.

# Chapter 4

# BLM FACES SEVERAL CHALLENGES TO FUTURE SALES UNDER FLTFA

BLM state and field office officials most frequently cited the availability of knowledgeable realty staff to conduct the sales as a challenge to raising revenue from FLTFA sales. These staff may not be available because they are working on activities that BLM has identified as a higher priority, such as reviewing and approving energy rights-of-way. We identified two additional issues hampering land sales activity under FLTFA. First, while BLM has identified land for sale in its land use plans, it has not made the sale of this land a priority during the first 7 years of the program. Furthermore, BLM has not set goals for FLTFA sales. Goals are an effective management tool for measuring and achieving results. Some BLM state offices reported that they have planned FLTFA sales through 2010, but BLM has no overall implementation strategy to generate funds to purchase inholdings, as mandated by FLTFA. Since BLM has not laid out a clear roadmap for FLTFA and did not make land sales a priority, it is difficult to determine if BLM took full advantage of the opportunities for generating revenue under the act. Second, BLM has revised some of its land use plans since 2000 and identified additional land for disposal. However, revenue from these potential sales is not eligible for the FLTFA account because the act only applies to land that was identified for disposal in a land use plan on or before the date of the act.

# BLM State and Field Officials Most Frequently Cited Availability of Knowledgeable Staff as a Challenge to Conducting FLTFA Sales

According to BLM state and field officials, they face five challenges to raising FLTFA revenue through sales. First, the most frequently identified is the availability of knowledgeable realty staff to conduct the sales. This challenge is followed, in order of frequency cited, by the time, cost, and complexity of the land sales process; external factors, such as public opposition to a sale; program and legal restrictions; and the land use planning process. Except for FLTFA-specific program and legal restrictions, the other challenges that BLM state and field offices cited are probably faced in many public land sales. The following provides examples of these challenges:

- *The availability of knowledgeable realty staff to conduct the sales.* BLM staff said realty staff must address higher priority work before land sales. For example, Colorado BLM staff said that processing rights-of-way for energy pipelines takes a huge amount of realty staff time, 100 percent in some field offices, and poses one of the top challenges to carrying out FLTFA sales in Colorado. In Idaho, staff also cited the lack of realty staffing, which is down 40 percent from 10 years ago. Adding to the staffing issue, the workload for energy-related uses in Idaho, such as approving rights-of-way for transmission lines, has doubled. Other offices cited turnover in staff and the lack of staff with training and experience to conduct sales.
- *Time, cost, and complexity of the sales process.* Much preparation must be completed before a property can be sold. For example, several offices cited the cost and length of the process that ensures a sale complies with environmental laws and regulations. In addition, obtaining clearances from experts related to cultural and natural resources on a proposed sale can be time-consuming. For example, in the sale of 396 acres by the Las Cruces District Office, officials said that the sale of the property was delayed by the discovery of a

significant cultural resource on the site. This was eventually resolved by BLM retaining the small portion of the original parcel containing the cultural resource.

- *External factors.* BLM officials cited such factors such as public opposition to a sale, market conditions, or lack of political support as challenges. For example, Colorado BLM officials said that they have faced strong local opposition to sales, and the El Centro Field Office staff in California cited the lack of demand for the land from buyers as a challenge. Some offices have experienced diminishing support of sales by local governments over the time a sale is prepared.

- *Program and legal restrictions.* The Arizona State Office staff and the Elko Field Office staff cited the sunset date of FLTFA, less than 3 years away, as a challenge because the sunset date may not allow enough time to complete many more sales. Other offices said the MOU provision requiring a portion of the land sale proceeds to be used by the three other agencies reduces BLM's incentive to do land sales because BLM keeps only 60 percent of the revenue. Another challenge to the disposal of land under FLTFA, especially in Nevada, has been the passage of land bills for Lincoln and White Pine counties.[17] The Lincoln County Land Act of 2000, as amended, directs BLM to deposit most of the proceeds from the disposal of not more than 103,328 acres into an account established by the act. The White Pine County Conservation, Recreation, and Development Act of 2006 directs BLM to deposit most of the proceeds from the disposal of not more than 45,000 acres into a similar account. In total, BLM staff estimate that, once mandated land use plan amendments are completed, the two acts will result in the removal of about 148,000 acres from FLTFA eligibility.

- *Land use planning.* Some offices cited problems with the land use plans. For example, the Idaho Falls District Office staff said that specific land for sale is hard to identify in old land use plans. Nevada's Elko Field Office staff said that some lands that could be offered for sale were not available because they were not designated in the land use plan at the time of FLTFA's enactment.

# MOST BLM STATES HAVE PLANNED FLTFA SALES THROUGH 2010, BUT BLM LACKS NATIONAL GOALS FOR THE PROGRAM

BLM state offices reported planning FLTFA sales through 2010, but BLM has not established national goals for FLTFA or emphasized sales.

## BLM Plans FLTFA Sales through 2010

In response to our request to the 10 BLM state offices participating in FLTFA, 8 reported planning 96 FLTFA sales totaling 25,406 acres through 2010. The other two state offices reported no planned sales. Of the 96 planned sales, 34 are planned as competitive, 6 as modified competitive, and 52 as direct sales; the sales methods for 4 sales are unknown. The BLM state offices did not report a fair market value for some of these planned sales. Table 3 provides information on planned FLTFA sales and appendix III provides a complete listing of the planned sales that BLM state offices reported.

### Table 3. BLM Reported Planned FLTFA Sales, through 2010

| BLM state office | Competitive | Direct | Modified competitive | Undetermined | Total planned sales | Total acres |
|---|---|---|---|---|---|---|
| Arizona | 3 | 2 | 1 | 2 | 8 | 2,640 |
| California | 1 | 4 | 0 | 0 | 5 | 251 |
| Colorado | 1 | 15 | 0 | 1 | 17 | 136 |
| Idaho | 2 | 13 | 1 | 0 | 16 | 4,242 |
| Montana | 0 | 0 | 0 | 0 | 0 | 0 |
| Nevada | 17 | 7 | 1 | 0 | 25 | 14,570 |
| New Mexico | 7 | 2 | 0 | 0 | 9 | 1,273 |
| Oregon/ Washington | 0 | 0 | 0 | 0 | 0 | 0 |
| Utah | 1 | 4 | 1 | 0 | 6 | 412 |
| Wyoming | 2 | 5 | 2 | 1 | 10 | 1,880 |
| **Total** | **34** | **52** | **6** | **4** | **96** | **25,406** |

Source: GAO analysis of information reported by BLM state offices.

Notes: Total acres column does not add due to rounding. An estimate of the expected total revenue from these sales was not available because all fair market values were not reported by the state offices.

Figure 5 shows an example of a planned sale—the "North Fork" parcel to be sold competitively in April 2008 by the field office in Las Cruces, New Mexico. This 167-acre parcel is on the eastern edge of Las Cruces across the street from residential subdivisions. BLM also plans to sell a similar adjacent 180-acre parcel at the same time. The field office reported that the purpose of these sales is to dispose of land that will serve important public objectives, including but not limited to, expansion of communities and economic development, which cannot be achieved prudently or feasibly on land other than public land.

Source: GAO.

Figure 5. BLM "North Fork" 167-Acre Parcel on the Eastern Edge of Las Cruces, New Mexico, Planned for Competitive Sale in an April 2008 Auction.

Although BLM offices plan sales, there is no assurance that these sales will occur. For fiscal years 2004 and 2005, BLM headquarters compiled a list of 100 planned sales under FLTFA from information the state offices provided. Because BLM headquarters did not know the status of these 100 planned sales, we followed up with the state offices and determined that 54 were actually completed. According to BLM's state office leads for these sales, 46 properties did not sell for several reasons, such as environmental concerns; external factors; the availability of staff; and the time, cost, and complexity of the sales. For example, Utah State Office officials said a

1,450-acre parcel near St. George did not sell because threatened and endangered species and cultural resource issues were identified. In Wyoming, state office staff said only one of four proposed sales occurred because of inadequate staffing and the competing priority to address oil and gas-related realty issues.

## BLM Has Not Established Goals or an Implementation Strategy for FLTFA Sales

BLM has established annual goals for the disposal of land through sales or other means [18]. For example, BLM's fiscal year 2008 budget justification contained a performance target to dispose of 11,500 acres and 30,000 acres of land in fiscal years 2007 and 2008, respectively. However, BLM has not established similar goals for FLTFA sales. For example, BLM's fiscal year 2007 annual work plan for the lands and realty function—which guides the activities to be completed in a given year—does not contain specific goals for FLTFA. Rather, it states that lands and realty staff should continue to hold land sales under FLTFA, especially in Nevada.

BLM did provide an estimate in its fiscal year 2008 budget justification for FLTFA revenue—$12 million in fiscal year 2007 and $50 million in 2008. However, BLM fell short of its estimate for fiscal year 2007; it reported generating only $0.7 million from sales and exchanges. Moreover, when we asked BLM headquarters staff for the basis of the fiscal year 2007 and 2008 revenue estimates, they said the estimates were based on professional judgment and that they had no supporting information [19].

Our interviews with state and field office staff confirmed that there are few goals for conducting FLTFA sales. According to 27 of the 28 state and field office officials we spoke with, BLM headquarters had not provided any goals; one state office said headquarters had emphasized getting their land disposal program up and running in their office. According to 18 of these 28 officials, their state and field office management had set no targets or goals for FLTFA land sales. Of the 10 that did mention such goals, 8 described the goal as a plan to sell specific parcels of land.

According to headquarters officials, BLM has tried to encourage FLTFA sales but is not pressuring field offices to conduct them, and there is no ongoing headquarters effort to oversee and manage sales because states are responsible for conducting their own sales programs. The realty managers explained that headquarters does not approve land sales but is aware of them through reviews of Federal Register notices of the sales. According to a headquarters official, BLM did not establish FLTFA goals because BLM lacked realty staff to conduct land sales and other impediments to sales generally, such as the lack of access, mineral leases, mining claims, threatened or endangered species habitat, floodplains, wetlands, cultural resources, hazardous materials, and title problems.

The establishment of goals is an effective management tool for measuring and achieving results. As we have reported in the past on management under the Government Performance and Results Act of 1993 [20], leading public sector organizations pursuing results-oriented management commonly took the following key steps:

- defined clear missions and desired outcomes,
- measured performance to gauge progress, and
- used performance information as a basis for decision making.

BLM has not fully implemented these steps in managing the FLTFA program to sell land designated for disposal in its land use plans. To measure BLM's success in generating revenue and disposing of land under FLTFA, actual performance would need to be compared with national sales goals for FLTFA. Without national goals for making these sales a priority, it is difficult for BLM to enhance the efficiency and effectiveness of federal land management as called for in FLTFA through the acquisition of inholdings and consolidation of public lands.

## FLTFA'S RESTRICTION ON LAND AVAILABLE FOR SALE REDUCES POTENTIAL REVENUE

FLTFA requires BLM to deposit the proceeds into the special FLTFA account from the sale or exchange of public land identified for disposal under approved land use plans in effect on the date of its enactment [21]. Other proceeds from land sales and exchanges are typically deposited into the U.S. Treasury's general account. Many of BLM's land use plans have been revised or have been proposed for revision since FLTFA's enactment, and additional lands have been identified for disposal. For example, BLM reported the Boise District Office in Idaho is currently planning a sale of 35 parcels. Five of the 35 parcels, with a total estimated value of $10.7

million, are not FLTFA eligible. Because of the land use plan restriction, revenue from these five sales would not benefit the FLTFA account when sold. While this restriction reduces the potential revenue that could be dedicated to purchasing inholdings and adjacent land containing exceptional resources under FLTFA, it does benefit the U.S. Treasury's general account.

Source: GAO.

Figure 6. A View of the West Mesa BLM Property in Las Cruces, New Mexico, That Will Be Added to Land Designated for Disposal in the Revised Land Use Plan.

According to 17 of the 28 BLM state and field realty staff we interviewed, their office has land available for disposal that is not designated in an FLTFA-eligible land use plan. For example, New Mexico state office officials said that a number of land use plan amendments completed or under development since FLTFA's enactment have identified land for disposal. They noted that the Las Cruces area land use plan is being amended to adjust to the new direction of the city's growth that has occurred since the last plan was prepared in 1993. According to BLM New Mexico staff, different or additional lands are expected to be designated for disposal in the amended plan. Figure 6 shows land on the west side of Las Cruces, New Mexico, that is expected to be designated for disposal in the forthcoming revision to accommodate the community's growth. Field office officials said that input from local governments and other interests have focused land sales growth in Las Cruces on the west side of the city in order to create a buffer for the Organ Mountains on the east side.

# Chapter 5

# AGENCIES HAVE PURCHASED FEW PARCELS WITH FLTFA REVENUE

Since the enactment of FLTFA 7 years ago, BLM reports that the four land management agencies have spent $13.3 million of the $95.7 million in FLTFA revenue—$10.1 million to acquire nine parcels of land and $3.2 million in administrative expenses for conducting FLTFA sales. Agencies spent the $10.9 million between August 2007 and January 2008 on the first land acquisitions completed under FLTFA using the secretarial discretion provided in the MOU. As of May 31, 2007, the agencies reported submitting eight acquisition nominations to state-level interagency teams for consideration. The New Mexico interagency team reported submitting six additional nominations as of July 1, 2007. None of these 14 nominations—valued at $71.9 million—has resulted in a completed acquisition. The state-level process has not yet resulted in acquisitions because of the time taken to complete interagency agreements and limited FLTFA funds available for acquisition outside of Nevada. Although Nevada has proposed five acquisitions, none have been completed. As for the remaining $3.2 million in expenditures, BLM reports spending these funds on administrative activities involved in preparing land for sale under FLTFA mostly between 2004 and 2007. BLM offices in Nevada spent $2.6 million of this total.

## Under a Secretarial Initiative,
## BLM Reports Agencies Spent $10.1 Million
## on the First Land Acquisitions
## 7 Years after FLTFA Was Enacted

No land acquisitions had occurred during the first 7 years of FLTFA. Because the state–level implementation process had not resulted in any acquisitions, BLM decided to jump-start the acquisition program and conduct purchases under secretarial discretion, as provided for in the MOU. In the spring of 2006, BLM headquarters officials solicited nominations from the FLTFA leads in each of the other three agencies. Most of the nominations agency officials provided were previously submitted for funding under LWCF. This secretarial initiative was approved by the Secretaries of Agriculture and of the Interior in May 2007.

The 2007 secretarial initiative provided $18 million in funding for 13 land acquisition projects, including 19 parcels of land located in seven states—Arizona, California, Colorado, Idaho, New Mexico, Oregon, and Wyoming. Specifically, the initiative consisted of 9,049 acres and included projects for each agency: six BLM projects for $10.15 million, two Fish and Wildlife Service projects for $1.75 million, two Forest Service projects for $3.5 million, and three Park Service projects for $2.6 million. Only 1 of the 19 parcels is an adjacent land; the rest are inholdings.

Since the initiative was approved, BLM reported a number of changes that the agencies made to the original list of land acquisition projects. For example, the total number of acres increased to 9,987 in a total of eight states. As of January 23, 2008, BLM reported that the agencies had wholly or partially completed 8 of the 13 approved acquisition projects. Specifically, the agencies spent $10.1 million between August 2007 and January 2008 to complete the acquisition of the first nine parcels under the secretarial initiative [22]. The acquisitions include 3,381 acres in seven states— Arizona, California, Idaho, Montana, New Mexico, Oregon, and Wyoming. See table 4 for a complete description of the current status of these projects.

**Table 4. Status of FLTFA Land Acquisition Projects Approved under the Secretarial Initiative, as of January 23, 2008**

| Agency | State | Federally designated area | Acres | FLTFA funding Status |
|---|---|---|---|---|
| **BLM** | | | | |
| | California | Coachella Valley Fringe-Toed Lizard Area of Critical Environmental Concern[a] | 321 | $850,000 Complete |
| | Colorado | Canyons of the Ancients National Monument | 469 | 500,000 Incomplete |
| | Idaho | Snake River Area of Critical Environmental Concern[b] | 1,674 | 4,700,000 Partially complete[c] |
| | New Mexico | La Cienega Area of Critical Environmental Concern El Camino Real de Tierra Adentro National Historic Trail | 178 | 2,200,000 Complete |
| | Oregon | Rogue National Wild and Scenic River | 32 | 600,000 Incomplete |
| | Wyoming | North Platte River Special Recreation Management Area California National Historic Trail Mormon Pioneer National Historic Trail Pony Express National Historic Trail Oregon National Historic Trail | 277 | 1,300,000 Complete |
| **Subtotal** | | | **2,951** | **$10,150,000** |
| **Fish and Wildlife Service** | | | | |
| | Montana | Red Rock Lakes National Wildlife Refuge | 2,159 | 1,425,000 Complete[d] |
| | Oregon | Siletz Bay National Wildlife Refuge | 42 | 325,000 Partially complete[e] |
| **Subtotal** | | | **2,201** | **$1,750,000** |

## Table 4. Continued

| Agency | State | Federally designated area | Acres | FLTFA funding Status |
|---|---|---|---|---|
| **Forest Service** | | | | |
| | Arizona | Tonto National Forest | 11 | 635,000 Complete |
| | California | Six Rivers National Forest Smith River National Recreation Area Goose Creek National Wild and Scenic River | 4,303 | 2,865,000 Incomplete |
| **Subtotal** | | | **4,314** | **$3,500,000** |
| **Park Service** | | | | |
| | Idaho | Nez Perce National Historic Park Nez Perce National Historic Trail | 510 | 200,000 Incomplete[f] |
| | New Mexico | Aztec Ruins National Monument | 10 | 200,000 Incomplete |
| | Wyoming | Grand Teton National Park | 1 | 2,200,000 Complete |
| **Subtotal** | | | **521** | **$2,600,000** |
| **Total** | | | **9,987[g]** | **$18,000,000** |

Source: GAO analysis of information provided by BLM headquarters.

[a]This is the only property in the secretarial initiative that is adjacent to federal land; the rest are inholdings.

[b] Five parcels are included in this project. In addition, BLM has added one 300-acre parcel valued at $500,000 to the list of parcels under this project. The BLM FLTFA program lead said this parcel was included because of its high resource value but added that the agencies will remain within the $18 million spending level approved by the Secretaries. Of these six parcels, five totaling 1,872 acres will be acquired through easement, and one totaling 102 acres will be acquired in fee.

[c]As of January 23, 2008, a 102-acre parcel had been acquired in fee and an easement had been acquired on a 300-acre parcel.

[d]BLM reports that an acquisition at the Arapaho National Wildlife Refuge originally included in this secretarial initiative failed due to expired options on the property. The Director of the Fish and Wildlife Service has substituted an acquisition at the Red Rock Lakes National Wildlife Refuge for the same amount of funding.

[e]Two parcels are included in this project.

[f]Two parcels, both for easements, are included in this project.

[g]Some of the acres acquired or planned for acquisition were funded in part by sources other than FLTFA: the Land and Water Conservation Fund and the Migratory Bird Conservation Fund.

Figure 7 shows part of the acquisition site within the La Cienega Area of Critical Environmental Concern. According to BLM, it selected this site for acquisition because (1) it is an archeologically rich area preserving ancient rock art and (2) the riparian cottonwood and willow forest that line the Santa Fe River and its La Cienega Creek tributary provide critical habitat for threatened and endangered wildlife, such as the bald eagle and southwest willow flycatcher. The final purchase price was $2.2 million.

Source: GAO.

Figure 7. Part of an Inholding in the BLM La Cienega Area of Critical Environmental Concern That Has Been Acquired with FLTFA Funding.

To fund the acquisitions in the secretarial initiative of $18 million, the BLM FLTFA program lead told us that the Secretaries approved the use of

- $14.5 million of the funds from the 20 percent of revenue available for acquisitions outside the state in which they were raised, and
- $3.5 million of the revenue not used for administrative activities supporting the land sales program [23].

# AGENCIES HAVE SUBMITTED NOMINATIONS UNDER THE STATE-LEVEL PROCESS, BUT NONE HAVE RESULTED IN A LAND ACQUISITION

In addition to the acquisitions in the secretarial initiative, the agencies have submitted 14 acquisition nominations valued at $71.1 million to state-level interagency teams for consideration, but not one has resulted in a completed acquisition. Of the $14.1 million in land acquisitions awaiting a secretarial decision, $13.7 million, or 97 percent, is for inholdings and $458,000 million—or 3 percent—is for adjacent land. Table 5 shows the data we gathered from BLM state offices on the status of the nominations that have been submitted.

The Nevada interagency team has submitted a total of five nominations for secretarial approval under FLTFA. It nominated two Forest Service acquisitions—a total of 705 acres valued at $4.76 million—in 2004. These were the first nominations submitted for secretarial approval under FLTFA. The Forest Service was unable to complete the purchases because of negotiating differences with the sellers. Of the remaining three Nevada nominations, one valued at $16 million was approved in November 2007, one valued at $10.6 million awaits approval, and one valued at $29 million has been withdrawn by the Nevada interagency team.

The recently approved Nevada nomination is for the Pine Creek State Park, an 80-acre inholding owned by the state of Nevada valued at $16 million (see fig. 8). BLM currently manages this inholding, which is located in BLM's Red Rock Canyon National Conservation Area. According to the BLM nomination package, BLM would like to acquire this property to meet the increasing recreational and educational needs of the park. BLM explains that the property has recreational value; cultural resources; riparian habitat; and habitat for the desert tortoise, currently a threatened and endangered species.

The nomination that was withdrawn by the Nevada interagency team is the 320-acre Winter's Ranch property, which is adjacent to the Humbolt-Toiyabe National Forest and several other properties acquired by BLM under SNPLMA. BLM's FLTFA program lead said the nomination of the parcel was withdrawn, in part because it is not adjacent to a federally designated area managed by BLM.

**Table 5. FLTFA Land Acquisition Nominations Reviewed by State-Level Interagency Teams, as of May 31, 2007[a]**

| Agency | State | Federally designated area | Inholding or adjacent land | Acres | Requested amount[b] | State-level interagency decision | If approved, status of secretarial approval, as of November 30, 2007 |
|---|---|---|---|---|---|---|---|
| **BLM** | | | | | | | |
| | Arizona | Hells Canyon Wilderness | Inholding | 640 | $3,000,000 | Approved | Not yet submitted |
| | California | Coachella Valley Fringe-Toed Lizard Area of Critical Environmental Concern | Adjacent | 301 | 458,000[c] | Approved | Pending |
| | Nevada | Red Rock Canyon National Conservation Area | Inholding | 80 | 16,015,000 | Approved | Approved |
| | Nevada | Humboldt-Toiyabe National Forest | Adjacent | 320 | 29,126,000 | Approved | Withdrawn |
| | New Mexico | Gila Lower Box Area of Critical Environmental Concern | Adjacent | 1,880 | 2,055,000 | Denied[d] | - |
| | New Mexico | Continental Divide National Scenic Trail | Inholding | 5,000 | 1,530,825 | Denied[e] | - |
| | New Mexico | Elk Springs Area of Critical Environmental Concern | Inholding | 2,280 | 1,810,000 | Approved | Pending |

## Table 5. Continued

| Agency | State | Federally designated area | Inholding or adjacent land | Acres | Requested amount[b] | State-level interagency decision | If approved, status of secretarial approval, as of November 30, 2007 |
|---|---|---|---|---|---|---|---|
| **Forest Service** | | | | | | | |
| | Nevada | Humboldt-Toiyabe National Forest | Inholding | 320 | 1,230,000 | Approved | Approved[f] |
| | Nevada | Humboldt-Toiyabe National Forest | Inholding | 385 | 3,530,000 | Approved | Approved[f] |
| | Nevada | Humboldt-Toiyabe National Forest | Inholding | 40 | 10,624,500 | Approved | Pending |
| | New Mexico | Cibola National Forest | Inholding | 160 | 160,000 | Denied[g] | - |
| | New Mexico | Santa Fe National Forest | Inholding | 160 | 660,000 | Approved | Pending |
| | New Mexico | Santa Fe National Forest | Inholding | 160 | 560,000 | Approved | Pending |
| | Wyoming | Bridger-Teton National Forest | Inholding | 40 | 388,600[h] | Pending | - |
| **Total** | | | | 11,766 | $71,147,925 | | |

Source: GAO analysis of information provided by BLM state offices.

[a] Although we requested information that had been updated as of May 31, 2007, the New Mexico interagency team provided information updated as of July 1, 2007. To provide the most current information, we are including the nominations in the July 1, 2007 data.

[b] Requested amount includes the estimated value of the parcel and, in some cases, administrative costs associated with the acquisition.

[c] The total value of this acquisition is estimated at $975,000. BLM plans to use LWCF funding to cover the remaining $517,000.

[d] This nomination was denied because it is not immediately adjacent to a federally designated area.

[e] While BLM nominated this parcel for acquisition, it was denied because the state interagency team determined this parcel is an inholding within a national forest.

[f] This acquisition was approved by the Secretaries but was ultimately terminated due to negotiating issues with the seller.

[g] This nomination was denied because it is a lower Forest Service priority.

[h] The total value of this acquisition is estimated at $412,600. The Forest Service plans to use other funding sources to cover the remaining $24,000.

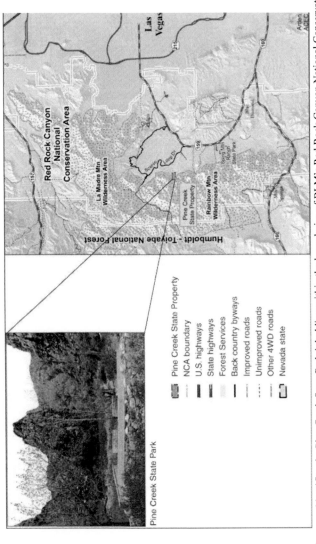

Note: The location map identifies the Pine Creek State Park inholding within the boundaries of BLM's Red Rock Canyon National Conservation Area located near Las Vegas, Nevada.

Figure 8. Photograph and Location of the 80-Acre Pine Creek State Park Inholding Approved as an FLTFA-Funded Acquisition at $16 Million.

In its nomination to acquire Winter's Ranch, the Carson City Field Office said this parcel provides critical habitat for shorebirds, water fowl, and other water-dependent species; offers unique recreational opportunities for the public; and an undisturbed view for area commuters and tourists. According to a Carson City Field Office official, three creeks run through this property and irrigate the land, making it possible to sustain wildlife habitat, such as raptors and migratory birds. The official said that about $20 million of the estimated $29 million value of the Winter's Ranch property is for the water rights to the property, and that if BLM did not obtain the water rights, other parties could acquire them and divert the water resources to other areas, such as developing communities near Reno. The Winter's Ranch parcel is shown in figure 9.

Source: GAO.

Figure 9. The 320-Acre Winter's Ranch FLTFA Acquisition Nomination in Nevada, Valued at $29 Million.

Over one-half of the state-level interagency teams—Colorado, Idaho, Montana, New Mexico, Oregon, and Utah—did not review any land acquisitions proposals between July 2000, when FLTFA was enacted, and May 2007. Furthermore, the Fish and Wildlife Service and the Park Service have yet to submit a nomination for review under the state-level interagency process. Fish and Wildlife Service and Park Service officials based in California said they lacked the FLTFA funding necessary to complete an acquisition and would have to wait until sufficient revenue were available to allow them to nominate an acquisition.

In examining the headquarters review and approval process, we found that the Land Transaction Facilitation Council established in the national MOU has never met. The BLM FLTFA program lead explained that, as a practical matter, it has not been necessary for this council to meet. Rather, in practice, acquisition nominations are forwarded to the BLM lead and then routed to his counterparts at the other three agencies for review.

Additional reviews are then conducted at the agency level and, ultimately, at the secretarial level.

## STATE-LEVEL PROCESS HAS NOT YET RESULTED IN ACQUISITIONS BECAUSE OF THE TIME TAKEN TO COMPLETE INTERAGENCY AGREEMENTS AND LIMITED FUNDS OUTSIDE OF NEVADA

Although the agencies envisioned it as the primary process for nominating land for acquisition under FLTFA, the state-level process established in the national MOU and state-level interagency agreements has yet to result in a completed land acquisition for two primary reasons. First, it has taken over 6 years for the four agencies to complete all interagency agreements—3 years for the agencies to complete a national MOU and an additional 3 years for the agencies to complete all state-level implementation agreements. Most agencies completed *Federal Register* notifications of their procedures to identify and set priorities for inholdings, as called for in the act, soon after state-level agreements were signed. Nevada was the first state to complete the implementation agreement in June 2004, and it published a *Federal Register* notice in August 2004. Utah was the last state to complete these actions, reaching an agreement in November 2006 and publishing its Federal Register notice in March 2007. Table 6 summarizes the completion of implementation agreements and the *Federal Register* publication for each state.

**Table 6. Completion of FLTFA Implementation Agreements and**
*Federal Register* **Notifications by State**

| State | Date implementation agreement signed | *Federal Register* publication date |
|---|---|---|
| Nevada | June 2004 | August 2004 |
| Montana | May 2005 | June 2006 |
| California | November 2005 | March 2006 |
| Colorado | January 2006 | August 2006 |
| Oregon/Washington | March 2006 | May 2006 |
| Arizona | May 2006 | July 2006 |
| Idaho | July 2006 | September 2006 |
| Wyoming | July 2006 | September 2006 |
| New Mexico | August 2006 | September 2006 |
| Utah | November 2006 | March 2007 |

Sources: FLTFA state-level interagency agreements and Federal Register notices.

BLM officials told us that completion of these agreements was delayed for a number of reasons, including attention to other priorities, difficulties coordinating the effort with four agencies, and lack of urgency due to limited revenue available for acquisitions.

Second, funds for acquisitions have been limited outside of Nevada. Because FLTFA requires that at least 80 percent of funds raised must be spent in the state in which they were raised and because 92 percent of funds have been raised in Nevada, the majority of funds must be spent on acquisitions in Nevada. However, as discussed earlier, no acquisitions have yet been completed in Nevada. Additional factors, such as the fact that about 92 percent of Nevada is already federally owned and that SNPLMA has provided additional resources for land acquisitions in Nevada, may have also contributed to the lack of a completed acquisition under FLTFA in Nevada.

Outside of Nevada, agencies have had little money to acquire land. Several agency officials, such as BLM state office officials in Utah and Oregon, told us that additional revenue needs to be generated under FLTFA for land acquisitions to occur. Moreover, Park Service and Forest Service officials in California told us they are waiting for adequate funding before they begin identifying and nominating acquisitions. The Forest Service official explained that the agency could not make significant purchases with their share of the FLTFA funds in California because of the high cost of real estate.

# BLM REPORTS SPENDING $3.2 MILLION ON FLTFA ADMINISTRATIVE ACTIVITIES

Between the time FLTFA was enacted and July 20, 2007, BLM reports spending $3.2 million on FLTFA administrative expenses to conduct land sales under the act. The three other agencies do not have land sale expenses under the program. The BLM Nevada offices spent 81 percent of the revenue, or $2.6 million. BLM offices in Arizona, California, New Mexico, and Oregon each spent over $100,000, and the remaining five states spent a combined total of less than $50,000. States with the most active sales programs generally spent the most FLTFA revenue. For example, Nevada field offices conducted 106 of the 265 total sales under FLTFA, or 40 percent of the sales. Table 7 summarizes administrative expenditures by state as reported by BLM's Division of Business Services.

**Table 7. BLM Reported Administrative Expenditures by State, July 25, 2000, through July 20, 2007**

| State | Expenditure amount |
|-------|-------------------:|
| Arizona | $103,636 |
| California | 123,119 |
| Colorado | 37,173 |
| Idaho | 652 |
| Montana | 0 |
| New Mexico | 171,712 |
| Nevada | 2,574,074 |
| Oregon/Washington | 145,930 |
| Utah | 7,319 |
| Wyoming | 4,022 |
| Other[a] | 4,319 |
| **Total** | **$3,171,956** |

Source: GAO analysis of BLM's Division of Business Services data.

[a] In addition to BLM state and field office expenditures, the Division of Business Services spent $2,595 and the BLM headquarters office spent $1,724—a total of $4,319.

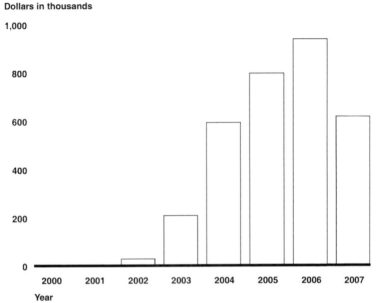

Source: GAO analysis of BLM's Division of Business Services data.

Figure 10. FLTFA Administrative Expenditures, July 25, 2000, through July 20, 2007.

### Table 8. FLTFA Administrative by Type, as of July 20, 2007

| Expenditure type | Amount | Percent of total |
|---|---|---|
| Personnel compensation and benefits[a] | $2,268,338[b] | 72 |
| Other services[c] | 829,948 | 26 |
| Printing and reproduction | 29,792 | 1 |
| Supplies and materials | 18,440 | 1 |
| Travel and transportation-personnel | 18,054 | 1 |
| Transportation of things | 7,093 | less than 1 |
| Rent, communications, and utilities | 292 | less than 1 |
| **Total** | **$3,171,957** | [d] |

Source: GAO analysis of BLM's Division of Business Services Data.

[a]Personnel compensation and benefits represent two expenditure types. Because they both involve payroll expenditures, we have combined them in this table.

[b]Of this amount, personnel compensation accounted for $1,819,397 and personnel benefits accounted for $448,941.

[c]Other services include expenditures such as appraisals and contracts for environmental and cultural activities.

[d]Due to rounding, the percentages do not add up to 100.

BLM spent little FLTFA revenue on the administrative costs of land sales during the first 3 years of the program. According to the BLM FLTFA program lead, there was little incentive for BLM to sell its land because the MOU was not in place. Spending has generally increased since then, with a spike in fiscal year 2006. Figure 10 shows FLTFA expenditures from its enactment to July 2007.

BLM's Division of Business Services tracks FLTFA expenditures through eight expenditure types. As table 8 shows, BLM offices spent 72 percent of FLTFA expenditures —about $2.3 million—on personnel compensation and benefits (e.g., staff to conduct sales).

*Chapter 6*

# AGENCIES FACE CHALLENGES IN COMPLETING ADDITIONAL ACQUISITIONS

BLM managers and we identified several challenges in completing additional acquisitions before FLTFA expires in 2010. BLM officials most commonly cited the time, cost, and complexity of the land acquisition process as a challenge to conducting acquisitions under FLTFA. We also found that the act's restriction on the use of funds outside of the state in which they were raised continues to limit acquisitions. Specifically, little revenue is available for acquisitions outside of Nevada. Furthermore, progress in acquiring priority land has been hampered by the agencies' weak performance in identifying inholdings and setting priorities for acquiring them, as required by the act. Finally, the agencies have yet to develop effective procedures to fully comply with the act and national MOU.

# BLM OFFICIALS MOST COMMONLY CITED THE TIME, COST, AND COMPLEXITY OF THE LAND ACQUISITION PROCESS AS A CHALLENGE, AMONG SEVERAL, TO COMPLETING ACQUISITIONS

BLM state and field officials from the 10 BLM state offices and 18 BLM field offices we interviewed most commonly cited the time, cost, and complexity of the land acquisition process as a challenge they face in completing land acquisitions. The other most commonly cited challenges were, in the order of frequency cited, (1) identifying a willing seller, (2) the availability of knowledgeable staff to conduct acquisitions, (3) the lack of funding to purchase land, (4) restrictions imposed by laws and regulations, and (5) public opposition to land acquisitions. Some of the challenges BLM state and field officials cited are likely typical of many federal land acquisitions. Because they have had little experience with FLTFA acquisitions, officials from the other three agencies had few comments about challenges. The following provides examples of each of these challenges:

- *Time, cost, and complexity of the land acquisition process.* To complete an acquisition under FLTFA, four agencies must work together to identify, nominate, and rank proposed acquisitions, which must then be approved by the two Secretaries. Officials at two field offices estimated the acquisition process takes about 2-1/2 to 3 years. BLM officials from the Wyoming State Office and the Las Cruces Field Office said that, with this length of time, BLM must either identify a very committed seller willing to wait to complete a transaction or obtain the assistance of a third party in completing an acquisition. A third party could help either by purchasing the land first, holding it, and then selling it to the government at a later date, or by negotiating with the seller an option to buy the land within a specified period. In terms of cost, some offices noted that they did not have the funding required to complete all of the work involved to prepare land acquisitions. In terms of complexity, a Utah State Office official said BLM has more control over the process for submitting land acquisitions under LWCF than FLTFA because

FLTFA requires four agencies in two departments to coordinate their efforts.

- *Identifying a willing seller.* Identifying a willing seller can be problematic because, among other things, the seller might have higher expectations of the property's value. For example, an Ely Field Office official explained that, because of currently high real estate values, sellers believe they can obtain higher prices from developers than from the federal government. Further, an Idaho State Office official said that it is difficult to find a seller willing to accept the appraised price and wait for the government to complete the purchase.

Even when land acquisition nominations are approved, they may not result in a purchase. For example, in 2004, under FLTFA, two approved acquisitions for inholdings within a national forest in Nevada were terminated. In one case, property values rose sharply during the nomination process and, in an effort to retain some of their land, the seller decided to reduce the acres for sale but maintain the price expectation. Furthermore, the landowner decided not to grant access through the parcel they were retaining to the Forest Service, thus eliminating the opportunity to secure access to an inaccessible area of the national forest. In the other case, during the course of the secretarial approval process, the landowner sold portions of the land included in the original transaction to another party, reducing the land available for the Forest Service to purchase. According to Forest Service officials, in both cases the purchase of the remaining parcels would not fulfill the original purpose of the acquisitions due to reductions in resource benefits. Therefore, the Forest Service terminated both projects. Similarly, the SNPLMA program in Nevada has had many terminated land acquisitions. Specifically, of the 116 land acquisition projects approved by the Secretary of the Interior from enactment in October 1998 through September 2007, 41 have been completed, 55 have been terminated, and 20 are pending. This represents a 47 percent termination rate. BLM did not report why these acquisitions were terminated.

- *Availability of knowledgeable staff to conduct acquisitions.* As is the case with selling federal land, BLM officials reported that they lack knowledgeable realty staff to conduct land acquisitions, as well as other BLM or department staff to conduct appraisals, surveys, and resource studies. Staff are occupied working on higher priority activities, particularly in the energy area.

- *Lack of funding to purchase land.* BLM officials in some states said they lack adequate funds to acquire land under FLTFA. For example, according to a field office official in Burns, Oregon, just one acquisition in a nearby conservation area would nearly drain that state's FLTFA account.

- *Restrictions imposed by laws and regulations.* BLM officials said that legal and other restrictions pose a challenge to acquiring land. BLM Arizona and Grand Junction, Colorado, officials said that some federally designated areas in their jurisdictions were established after the date of FLTFA's enactment, making the land within them ineligible for acquisition under the act. BLM New Mexico officials said that FLTFA's requirement that land be inholdings or adjacent land is too limiting and argued that the law generally should allow for the acquisition of land that has high resource values. In terms of regulations, BLM Carson City Field Office officials told us that the requirements they must follow regarding the processing of title, survey, and hazardous materials issues pose a challenge to conducting acquisitions.

- *Public opposition to land acquisitions.* According to BLM officials from the Elko and Ely Field Offices in Nevada, the public does not support the federal government's acquisition of federal land in their areas, arguing that the government already owns a high percentage of land and that such acquisitions result in the removal of land from the local tax base.

## COMPLIANCE WITH SPECIFIC PROVISIONS IN FLTFA CONTINUE TO POSE CHALLENGES TO FUTURE ACQUISITIONS

FLTFA's restriction on the use of funds outside of the state in which they were raised continues to limit acquisitions. Specifically, as mentioned earlier, little revenue is available for acquisitions outside of Nevada.

Furthermore, the Secretaries of Agriculture and of the Interior have given only minimal attention to developing a procedure specific to FLTFA for identifying inholdings and adjacent land and setting priorities for acquiring them, as required by the act. According to BLM's Assistant Director for Minerals, Realty, and Resource Protection, the four agencies met this requirement through their 2003 MOU. The official explained that the MOU establishes "a program for identification of eligible lands or interests in lands, and a process for prioritizing such lands or interests for acquisition." However, we found that the MOU only restates the basic statutory language for this requirement and states that the Secretaries are to establish a mechanism for identifying and setting priorities for acquiring inholdings. We found no such mechanism or procedure at the national level. While the state-level agreements do establish a process for reviewing proposed acquisitions, six minimally elaborate and three do not elaborate on the basic FLTFA criteria: the date the inholding was established, the extent to which the acquisition will facilitate management efficiency [24], and other criteria the Secretaries consider appropriate. One exception to this is the Nevada state-level agreement. Because the agencies involved in SNPLMA had already developed an interagency agreement to implement that act, they modified that agreement to include FLTFA. The Nevada agreement is generally more detailed than other state agreements and includes more criteria for considering land acquisitions because of the differences between the SNPLMA and FLTFA land acquisition authorities. Also, unlike the other state agreements, the Nevada agreement uses a quantitative system to rank acquisitions. Table 9 is a summary of criteria each state-level agreement includes beyond the FLTFA criteria for acquisition nominations.

When the agencies decided in 2006 to use the Secretaries' discretionary authority to make the initial FLTFA acquisitions, officials from all four agencies told us they generally relied on acquisition proposals previously identified for LWCF funding to quickly identify the parcels to acquire [25]. The agencies have systems to identify and set priorities for land acquisitions under LWCF. These existing systems could serve as a basis for systematically identifying and ranking FLTFA-eligible land for future acquisitions.

**Table 9. Additional Criteria Contained in FLTFA State-Level Agreements beyond Those Criteria Established under the Act**

| State agreement | Availability of funding | Completeness of nomination package | Local support | Criteria Agency prioritization | Contributes toward the preservation of a specially designated species | Estimated post-acquisition management costs | Other |
|---|---|---|---|---|---|---|---|
| Arizona | X | X | | | | | |
| California | X | X | X | X | | | |
| Colorado | X | X | | X | | | |
| Idaho | X | X | X | X | | | |
| Montana | | | | | | | |
| Nevada | | | X | X | X | X | X[a] |
| New Mexico | | | | | | | |
| Oregon/ Washington | X | X | | X | | | |
| Utah | X | X | | X | | | |
| Wyoming | X | X | | X | | | |

Source: FLTFA state-level implementation agreements.

[a]Three additional criteria are included in the Nevada state-level implementation agreement: (1) preserves a significant natural, aesthetic or scientific feature; (2) preserves significant historic, paleontological, or cultural site; and (3) enhances recreational opportunities or improves public access to recreational opportunities.

# THE AGENCIES HAVE YET TO ESTABLISH EFFECTIVE PROCEDURES TO FULLY COMPLY WITH FLTFA AND MOU PROVISIONS

With respect to FLTFA, the agencies—and primarily BLM, as the manager of the FLTFA account—have not established a procedure to track the act's requirement that at least 80 percent of funds allocated toward the purchase of land within each state must be used to purchase inholdings and that up to 20 percent may be used to purchase adjacent land [26]. The BLM FLTFA program lead said BLM considers this requirement when making land acquisition decisions but has not established a system to track it. The program lead noted that the requirement to use 80 percent for inholdings is hard to track, as the act is written, because the acquisition proposals are submitted in a piecemeal fashion.

With respect to the national MOU, BLM has not established a procedure to track agreed-upon fund allocations—60 percent for BLM, 20 percent for the Forest Service, and 10 percent each for the Fish and Wildlife Service and the Park Service [27]. The BLM FLTFA program lead told us the MOU allocations should be treated as a target or a goal on a national basis and they do not apply within a state. However, officials from the BLM Division of Business Services and BLM's Budget Office told us there is no mechanism to track these allocations and were unable to tell us whether the allocations should be followed at the state or national level. Knowing whether the MOU fund allocations are set at the state or national level is important because allocations that apply nationally provide more flexibility than allocations at the state level. While BLM did not track the allocations, most state-level interagency agreements provide guidance on consideration of nominations that exceed the established allocations and some BLM state office officials we spoke with were mindful of these allocation targets. For example, in California, the interagency team had agreed to "lend" BLM the funds from their allocations for a proposed BLM acquisition because they themselves could not effectively use the small portions of funding allocated to them. In contrast, in Oregon, BLM officials said they had not considered such an arrangement. The BLM FLTFA program lead said the funding decisions made by the Secretaries will be tracked and further information will be provided to the state-level interagency teams to clear up any misunderstanding of the requirement [28].

*Chapter 7*

# CONCLUSIONS

Congress anticipated that FLTFA would increase the efficiency and effectiveness of federal land management by allowing the four agencies to use certain land sales revenue without further appropriation to acquire priority land. Seven years later, BLM has not taken full advantage of the opportunity FLTFA offered. BLM has raised most of the funds for the FLTFA account with land sales in just one state, and it and the other land management agencies have made limited progress in acquiring inholdings and adjacent land with exceptional resources. Because there are less than 3 years remaining until FLTFA expires and a significant amount of time is needed to complete both sales and acquisitions, relatively little time remains to improve the implementation of FLTFA.

We recognize that a number of challenges have prevented BLM from completing many sales in most states, which limits the number of possible acquisitions. Many of the challenges that BLM cited are likely faced in many public land sales, as FLTFA did not change the land sales process. However, we believe that BLM's failure to set goals for FLTFA sales and develop a sales implementation strategy limits the agency's ability to raise revenue for acquisitions. Without goals and a strategy to achieve them, BLM field offices do not have direction for FLTFA sales. Moreover, the lack of goals makes it difficult to determine the extent of BLM's progress in disposing of unneeded lands to raise funds for acquisitions.

As with sales, progress in acquiring priority land has been hampered by weak agency performance in developing an effective mechanism to identify potential land acquisitions and set priorities for inholdings and adjacent land with exceptional resources, which FLTFA requires. Without such a mechanism, it is difficult to assess whether the agencies are acquiring the most significant inholdings and, thus, enabling them to more effectively and efficiently manage federal lands. Although the agencies do have systems to identify and set priorities for land acquisitions under LWCF that could potentially be adapted for the FLTFA acquisitions as well, they have not done so. Moreover, because the agencies have not tracked the amounts spent on inholdings and agency allocations, they cannot ensure compliance with the act or full implementation of the MOU.

As Congress considers the Administration's proposal to amend and reauthorize FLTFA, it may wish to reconsider the act's requirements that eligible lands are only those designated in the land use plans at the time FLTFA was enacted and that most FLTFA revenue raised must be spent in that state. Adjusting the eligibility of land use plans, as the Administration has proposed, could provide additional resources for land acquisitions under FLTFA. In addition, providing the agencies with more flexibility over the use of funds may allow them to acquire the most desirable land nationwide.

*Chapter 8*

# MATTERS FOR CONGRESSIONAL CONSIDERATION

If Congress decides to reauthorize FLTFA in 2010, it may wish to consider revising the following provisions to better achieve the goals of the act:

- *FLTFA limits eligible land sales to those lands identified in land use plans in effect as of July 25, 2000.* This provision excludes more recently identified land available for disposal, thereby reducing opportunities for raising additional revenue for land acquisition.
- *The requirement that agencies spend the majority of funds raised from eligible sales for acquisitions in the same state.* This provision makes it difficult for agencies to acquire more desirable land in states that have generated little revenue.

# OBJECTIVES, SCOPE, AND METHODOLOGY

With the Federal Land Transaction Facilitation Act of 2000 (FLTFA) set to expire in July 2010, we were asked to (1) determine the extent to which the Bureau of Land Management (BLM) has generated revenue for the FLTFA program, (2) identify challenges BLM faces in conducting future sales, (3) determine the extent to which the agencies have spent funds under FLTFA, and (4) identify challenges the agencies face in conducting future acquisitions. We also assessed the reliability of data BLM provided on revenue generated and on expenditures to date under FLTFA.

For all four objectives, we reviewed FLTFA, other applicable laws, regulations, and agency guidance. We interviewed the FLTFA program leads at the headquarters offices for BLM, the Fish and Wildlife Service, and the Park Service within the U.S. Department of the Interior, and the Forest Service within the U.S. Department of Agriculture on program status, goals, and management oversight for the program. To understand BLM's interpretation of key provisions of the act, we interviewed officials with Interior's Office of the Assistant Secretary for Land and Minerals Management and Office of the Solicitor and, in some cases, requested the department's views on these provisions in writing.

To determine the extent to which BLM has generated and expended FLTFA program revenue, we obtained and analyzed data from BLM's Division of Business Services on program revenue and visited Division of Business Services accounting officials in Lakewood, Colorado, to discuss the management of the FLTFA account. Using information provided by the Division of Business Services and information we obtained from the *Federal*

*Register*, we prepared summary information on completed sales by state and asked the 10 BLM state office officials responsible for the FLTFA program in their state to verify and update that information. As part of the request to state offices, we obtained data on planned FLTFA land sales and completed and planned acquisitions through 2010. We subjected the data provided by the field offices to electronic and logic testing and followed up with the field contacts regarding questions. With regard to acquisitions, we reviewed available documentation for land acquisition proposals considered by the 10 FLTFA interagency teams at the state level, agency headquarters, and the Secretaries of Agriculture and of the Interior. During our visits to selected BLM state offices (California, Nevada, New Mexico, and Oregon) and field offices (Carson City, Nevada, and Las Cruces, New Mexico), we interviewed officials and visited planned land acquisition sites to learn about the land acquisition process. During these visits we also interviewed selected officials with the Fish and Wildlife Service, the Forest Service, and the Park Service to learn about their experience in drafting state-level interagency agreements and with implementing land acquisitions under FLTFA. To assess the reliability of data provided by the Division of Business Services on revenue and expenditures, we interviewed staff responsible for compiling and reporting the data at the Division of Business Services and at the state office and field locations visited. We examined reports of this data from BLM's financial systems and related guidance and sought documentation on selected entries into the system.

To determine whether BLM has sufficient internal controls over FLTFA receipts and expenditures, we interviewed officials at the bureau's Division of Business Services and obtained, reviewed, and assessed the system of internal controls for the U.S. Treasury account established under FLTFA, including management's written policies and procedures, as well as control activities over collections, expenditures, and the records for these transactions. We also reviewed documentation for a non-probability sample of 7 nonlabor FLTFA expenditures totaling $54,967 that were charged by the Las Cruces Field Office to ensure proper documentation. As of July 20, 2007, BLM offices had made a total of 15,706 expenditure transactions—858 nonlabor and 14,848 labor—nationwide. The seven we chose included expenditures for appraisals and cultural evaluations on properties being prepared for sale under FLTFA. We chose these transactions because they were the largest ones and included a single vendor. We also chose one expenditure made on a charge card because it was slightly less than a reporting limit. We checked to ensure that documentation for these expenditures included (1) an agreement or contract between BLM and the entity to have specific work completed, (2) an invoice detailing work performed, and (3) evidence of BLM supervisory approval to pay for such services. After our review of the internal control policies and procedures, testing and verification of data on revenue, and obtaining documentation of the selected expenditures, the revenue and expenditure data was considered sufficiently reliable for our book.

To identify challenges to conducting land sales and acquisitions, we reviewed the FLTFA national memorandum of understanding, state-level interagency agreements, and documentation of headquarters and state-level interagency team activities to learn about the policies and procedures established for the implementation of FLTFA. We conducted semistructured interviews using a web-based protocol with (1) the 10 BLM state officials responsible for the FLTFA program in their state—Arizona, California, Colorado, Idaho, Montana, Nevada, New Mexico, Oregon/Washington, Utah, and Wyoming; (2) officials at the seven BLM field offices that have raised 97 percent of the FLTFA revenue (as shown in table 10); and (3) a nongeneralizable sample of 11 of the 137 remaining

BLM field offices that had not conducted a competitive sale under FLTFA as of May 31, 2007 (as shown in table 11). From the field offices with no competitive sales, we choose at least one office from each of the ten state offices under FLTFA and we considered the proximity of lands managed by field offices to urban areas. For California, we selected two additional field offices—Palm Springs and Eagle Lake. We chose the Palm Springs Field Office because it planned a major sale during our review and we chose the Eagle Lake Field Office because, although it is located in California, it manages some land in Nevada and has had no competitive sales. Because all of the Nevada field offices have had competitive sales and four Nevada offices were among the high revenue offices selected, we decided to select the Eagle Lake office. To analyze the narrative responses to some of the semistructured interview questions, we used the web-based system to perform content analyses of select open-ended responses. To conduct the content analyses to develop statistics on agreement among the answers, two reviewers per question collaborated on developing content categories based on survey responses and independently assessed and coded each survey response into those categories. Intercoder reliability (agreement) statistics were electronically generated in the coding process, and agreement on all categories were 90 percent or above. Coding disagreements were resolved through reviewer discussion. In addition, analyses of the closed-ended responses were produced with statistical software.

**Table 10. The Seven BLM Field Offices Selected That Have Generated
97 Percent of FLTFA Revenue, as of May 31, 2007**

| State | BLM office |
|---|---|
| Nevada | Carson City Field Office |
| Nevada | Elko Field Office |
| Nevada | Ely Field Office |
| Nevada | Las Vegas Field Office |
| New Mexico | Las Cruces District Office |
| Oregon | Burns District Office |
| Wyoming | Rock Springs Field Office |

Source: GAO analysis of BLM Division of Business Services revenue data.

**Table 11. The Eleven BLM Field Offices Selected That Had Not
Conducted a Competitive Sale under FLTFA, as of May 31, 2007**

| State | BLM office |
|---|---|
| Arizona | Lower Sonoran Field Office |
| California | Eagle Lake Field Office |
| California | El Centro Field Office |
| California | Palm Springs Field Office |
| Colorado | Grand Junction Field Office |
| Idaho | Idaho Falls District Office |
| Montana | Lewistown Field Office |
| New Mexico | Farmington Field Office |
| Oregon | Medford District Office |
| Utah | St. George Field Office |
| Wyoming | Casper Field Office |

Source: GAO analysis of BLM Division of Business Services revenue data and other factors.

We also interviewed a range of officials about the land acquisition process. These officials included FLTFA program leads at each agency's headquarters and selected state or regional-level contacts with each agency, as well as officials from third-party organizations involved with the land acquisition process, such as The Nature Conservancy and The Trust for Public Land.

We performed our work between November 2006 and February 2008 in accordance with generally accepted government auditing standards. Those standards require that we plan and perform the audit to obtain sufficient, appropriate evidence to provide a reasonable basis for our findings and conclusions based on our audit objectives. We believe that the evidence obtained provides a reasonable basis for our findings and conclusions based on our audit objectives.

# COMPLETED FLTFA LAND SALES, THROUGH MAY 2007

**Appendix II.**

| State and office | Customer | Transaction date | Amount collected | Acreage | Fair market value | Sale price | Disposal method |
|---|---|---|---|---|---|---|---|
| **Arizona** | | | | | | | |
| Arizona Strip District Office | Betty Foster/Marble Canyon Co. Inc. | 06/19/2002 | $21,600 | 22.77 | $21,600 | $21,600 | Direct |
| [a] | Joe W. Wright | 02/27/2002 | 32,502 | 5 | 32,500 | 32,502 | Competitive |
| **California** | | | | | | | |
| Eagle Lake Field Office | Pitchfork Cattle Co. | 02/04/2003 | 8,000 | 40 | 8,000 | 8,000 | Direct |
| Folsom Field Office | Michael Spence | 05/20/2002 | 10,000 | 3.75 | 10,000 | 10,000 | Direct |
| Folsom Field Office | Joseph Parisi | 12/03/2003 | 7,660 | 1.86 | [a] | [a] | Direct |
| Folsom Field Office | Fidelity National Title/Joseph Parisi | 12/03/2003 | 25,000 | 4.06 | 25,000 | 25,000 | Direct |
| Folsom Field Office | Isak Hanson | 12/06/2002 | 5,000 | 1.17 | 5,000 | 5,000 | Direct |
| Folsom Field Office | Operating Engineers Local Union No 3/ Frank Rocha | 01/12/2006 | 15,050 | 1.59 | 15,000 | 15,000 | Direct |
| Folsom Field Office | Big Oak Flat 1st Baptist Church | 05/01/2003 | 2,050 | 1.89 | 2,000 | 2,050 | Direct |
| Redding Field Office | Richard Dews | 10/07/2002 | 7,250 | 7.36 | 7,250 | 7,250 | Direct |
| Redding Field Office | Trinity County | 05/14/2002 | 135,000 | 123.28 | 135,000 | 135,000 | Direct |
| Redding Field Office | Department Of General Services | 05/09/2003 | 20,000 | 30.14 | 20,000 | 20,000 | Direct |

**Appendix II. Continued**

| State and office | Customer | Transaction date | Amount collected | Acreage | Fair market value | Sale price | Disposal method |
|---|---|---|---|---|---|---|---|
| **Colorado** | | | | | | | |
| Kremmling Field Office | Peter Van Veen | 10/18/2005 | 268,100 | 41.26 | 268,100 | 268,100 | Modified competitive |
| Kremmling Field Office | Ray Miller | 06/23/2006 | 67,500 | 2.65 | 67,500 | 67,500 | Competitive |
| Kremmling Field Office | Stephen Rector | 10/18/2005 | 16,000 | 3.2 | 12,000 | 16,000 | Modified competitive |
| Royal Gorge Field Office | Gary Munson | 10/07/2002 | 19,300 | 3.17 | 19,300 | 19,300 | Direct |
| Royal Gorge Field Office | Debora Evangelista | 06/27/2002 | 38,600 | 2.32 | 38,600 | 38,600 | Direct |
| Royal Gorge Field Office | Tom Cardinale | 04/11/2002 | 2,610 | 0.6 | 2,610 | 2,610 | Direct |
| Royal Gorge Field Office | Lenore Seiler | 10/07/2002 | 9,680 | 1.21 | 9,680 | 9,680 | Direct |
| Royal Gorge Field Office | Cripple Creek & Victor Gold Mining Company | 11/18/2003 | 21,361 | 5.76 | 21,361 | 21,361 | Competitive |
| Royal Gorge Field Office | Cripple Creek & Victor Gold Mining Company | 11/18/2003 | 24,622 | 6.14 | 24,622 | 24,622 | Competitive |
| Royal Gorge Field Office | Cripple Creek & Victor Gold Mining Company | 11/18/2003 | 11,405 | 1.51 | 11,405 | 11,405 | Competitive |
| Royal Gorge Field Office | Cripple Creek & Victor Gold Mining Company | 11/18/2003 | 18,505 | 5.45 | 18,505 | 18,505 | Competitive |
| Royal Gorge Field Office | Cripple Creek & Victor Gold Mining Company | 11/18/2003 | 10,538 | 3.05 | 10,538 | 10,538 | Competitive |
| Royal Gorge Field Office | Cripple Creek & Victor Gold Mining Company | 11/18/2003 | 50,720 | 10.42 | 50,720 | 50,720 | Competitive |
| Royal Gorge Field Office | County of Boulder | 06/19/2002 | 20,540 | 1.95 | 20,540 | 20,540 | Direct |

# Appendix II. Continued

| State and office | Customer | Transaction date | Amount collected | Acreage | Fair market value | Sale price | Disposal method |
|---|---|---|---|---|---|---|---|
| San Juan Public Land Center | Daren and Georgia Hillery | 11/18/2002 | 6,400 | 3.88 | 6,400 | 6,400 | Direct |
| San Luis Valley Public Land Center | Mountain Valley Lumber | 05/09/2007 | 31,500 | 28.63 | 31,500 | 31,500 | Direct |
| San Luis Valley Public Land Center | Hal Wilsohn | 08/27/2006 | 2,250 | 3.21 | 2,250 | 2,250 | Direct |
| White River Field Office | Taylor Temples | 08/01/2002 | 25,750 | 2.49 | 25,750 | 25,750 | Direct |
| White River Field Office | Walter Powell | 06/13/2002 | 2,500 | 3.35 | 2,500 | 2,500 | Direct |
| White River Field Office | Gary Staley | 06/17/2002 | 11,775 | 7.85 | 11,775 | 11,775 | Direct |
| White River Field Office | Mark Slawson | 06/04/2002 | 4,500 | 7.5 | 4,500 | 4,500 | Direct |
| White River Field Office | Chris Halandras | 05/22/2002 | 40,000 | 10.47 | 40,000 | 40,000 | Direct |
| White River Field Office | Cross Slash 4 Ranch | 11/18/2002 | 10,000 | 2.52 | 10,000 | 10,000 | Direct |
| White River Field Office | Big Mountain Ranch, LLC | 06/06/2002 | 160,000 | 80 | 160,000 | 160,000 | Direct |
| White River Field Office | Howard Cooper | 04/21/2006 | 65,700 | 4.84 | 65,700 | 65,700 | Direct |
| **Idaho** | | | | | | | |
| Boise District Office | Ada County | 04/24/2002 | 116,000 | 80 | 116,000 | 116,000 | Direct |
| Boise District Office | Owyhee County | 06/06/2005 | 9,000 | 30 | 9,000 | 9,000 | Direct |
| Challis Field Office | David W. Baker | 05/18/2006 | 9,600 | 5 | 9,600 | 9,600 | Direct |
| Challis Field Office | David W. Baker | 05/18/2006 | 5,000 | 3.09 | 5,000 | 5,000 | Direct |
| Challis Field Office | Gerald Nelson | 04/29/2002 | 24,600 | 49.2 | 24,600 | 24,600 | Direct |

**Appendix II. Continued**

| State and office | Customer | Transaction date | Amount collected | Acreage | Fair market value | Sale price | Disposal method |
|---|---|---|---|---|---|---|---|
| Challis Field Office | Sydney Dowton | 02/26/2002 | 5,940 | 31.27 | 5,940 | 5,940 | Direct |
| Challis Field Office | Firstfruits Foundation | 04/02/2007 | 2,600 | 0.75 | 2,600 | 2,600 | Direct |
| a | Susan Davis | 08/28/2003 | 2,200 | 0.96 | 2,200 | 2,200 | Direct |
| Upper Snake Field Office | Byron and Teresa Blakely | 04/10/2006 | 5,800 | 5.81 | 5,800 | 5,800 | Direct |
| **Montana** | | | | | | | |
| South Dakota Field Office | Wharf Resources | 04/29/2005 | 300 | 4.05 | 300 | 300 | Direct |
| South Dakota Field Office | Golden Reward Mining Co. | 01/12/2005 | 700 | 8.93 | 700 | 700 | Direct |
| South Dakota Field Office | S and J Cattle | 11/28/2006 | 58,000 | 40 | 26,000 | 58,000 | Competitive |
| **Nevada** | | | | | | | |
| Battle Mountain Field Office | Homestake Mining Co. | 06/27/2003 | 70,450 | 351.24 | 70,450 | 70,450 | Direct |
| Carson City Field Office | Carson Auto Mall, LLC | 01/29/2004 | 14,600,000 | 146 | 6,500,000 | 14,600,000 | Competitive |
| Carson City Field Office | John Serpa | 04/24/2006 | 8,400,000 | 106.25 | 6,400,000 | 8,400,000 | Competitive |
| Carson City Field Office | Raymond Sidney | 07/17/2006 | 16,100,000 | 100 | 10,000,000 | 16,100,000 | Competitive |
| Carson City Field Office | Jacob and Arezou Saeedi | 09/08/2005 | 608,000 | 30 | 297,000 | 608,000 | Competitive |
| Elko Field Office | Mike Gerber | 09/28/2004 | 16,500 | 40 | 4,600 | 16,500 | Competitive |
| Elko Field Office | Rabab Mardini | 09/20/2004 | 62,000 | 249.4 | 59,000 | 62,000 | Competitive |
| Elko Field Office | Elko Land and Livestock | 10/16/2006 | 36,000 | 79.5 | 25,000 | 36,000 | Competitive |

**Appendix II. Continued**

| State and office | Customer | Transaction date | Amount collected | Acreage | Fair market value | Sale price | Disposal method |
|---|---|---|---|---|---|---|---|
| Elko Field Office | West Wendover Project, LLC | 09/26/2006 | 56,000 | 320 | 100,000 | 280,000 | Competitive |
| Elko Field Office | NRLL East, LLC | 09/26/2006 | 12,000 | 40 | 9,000 | 12,000 | Competitive |
| Elko Field Office | Rabab Mardini | 11/09/2006 | 175,000 | 80 | 65,000 | 175,000 | Competitive |
| Elko Field Office | NRLL East, LLC | 09/26/2006 | 20,000 | 120 | 20,000 | 20,000 | Competitive |
| Ely Field Office | Steven Klomp | 06/19/2003 | 15,000 | 1.25 | 5,000 | 15,000 | Direct |
| Ely Field Office | Mt. Wheeler Power, Inc. | 07/23/2002 | 11,000 | 5 | 11,000 | 11,000 | Competitive |
| Ely Field Office | City of Caliente | 12/30/2002 | 150,000 | 105.21 | 150,000 | 150,000 | Direct |
| Las Vegas Field Office | Jon and Robin I Hadley | 06/16/2005 | 25,000 | 1.25 | 69,000 | 125,000 | Competitive |
| Las Vegas Field Office | Adriana Velez | 12/05/2005 | 104,000 | 1.25 | 69,000 | 125,000 | Competitive |
| Las Vegas Field Office | Varinder Singh | 12/12/2005 | 125,000 | 1.25 | 69,000 | 165,000 | Competitive |
| Las Vegas Field Office | Nye County, Nevada | 03/03/2003 | 47,900 | 10.66 | 47,900 | 47,900 | Direct |
| Las Vegas Field Office | Scott Gaughan | 08/02/2005 | 400,000 | 80 | 290,000 | 400,000 | Competitive |
| Las Vegas Field Office | Scott Gaughan | 02/03/2005 | 200,000 | 160 | 16,000 | 200,000 | Competitive |
| Las Vegas Field Office | Scott Gaughan | 02/03/2005 | 225,000 | 119.16 | 10,000 | 225,000 | Competitive |
| Las Vegas Field Office | State of Nevada, Division of State Lands | 05/07/2004 | 123,944 | 22.5 | 124,000 | 124,000 | Direct |
| Las Vegas Field Office | D.J. Laughlin | 07/20/2006 | 3,450,000 | 20 | 2,280,000 | 3,450,000 | Competitive |
| Las Vegas Field Office | Hamid Michael Mahban | 12/12/2005 | 425,000 | 5 | 275,000 | 425,000 | Competitive |
| Las Vegas Field Office | Darrell Hammer | 06/16/2005 | 420,000 | 5 | 275,000 | 420,000 | Competitive |
| Las Vegas Field Office | Hamid Michael Mahban | 12/12/2005 | 200,000 | 2.5 | 138,000 | 200,000 | Competitive |

**Appendix II. Continued**

| State and office | Customer | Transaction date | Amount collected | Acreage | Fair market value | Sale price | Disposal method |
|---|---|---|---|---|---|---|---|
| Las Vegas Field Office | The Crescent Group, LLC | 12/09/2005 | 200,000 | 2.5 | 138,000 | 200,000 | Competitive |
| Las Vegas Field Office | John W. Luther | 10/20/2005 | 195,000 | 2.5 | 138,000 | 195,000 | Competitive |
| Las Vegas Field Office | Patrick Paoli | 12/08/2005 | 195,000 | 2.5 | 138,000 | 195,000 | Competitive |
| Las Vegas Field Office | Philip Davis | 12/12/2005 | 195,000 | 2.5 | 138,000 | 195,000 | Competitive |
| Las Vegas Field Office | Philip Davis | 12/12/2005 | 210,000 | 2.5 | 138,000 | 210,000 | Competitive |
| Las Vegas Field Office | Philip Davis | 12/12/2005 | 260,000 | 2.5 | 138,000 | 260,000 | Competitive |
| Las Vegas Field Office | Philip Davis | 12/12/2005 | 275,000 | 2.5 | 138,000 | 275,000 | Competitive |
| Las Vegas Field Office | Philip Davis | 12/12/2005 | 495,000 | 5 | 275,000 | 495,000 | Competitive |
| Las Vegas Field Office | Philip Davis | 12/12/2005 | 600,000 | 5 | 275,000 | 600,000 | Competitive |
| Las Vegas Field Office | The Crescent Group, LLC | 12/09/2005 | 130,000 | 1.25 | 69,000 | 130,000 | Competitive |
| Las Vegas Field Office | John Koster | 06/16/2005 | 135,000 | 1.25 | 69,000 | 135,000 | Competitive |
| Las Vegas Field Office | Hardy Properties, LLC | 12/08/2005 | 120,000 | 1.25 | 69,000 | 120,000 | Competitive |
| Las Vegas Field Office | Robert Dale Beck | 06/16/2005 | 125,000 | 1.25 | 69,000 | 125,000 | Competitive |
| Las Vegas Field Office | Rajwinder Dhaliwal | 12/06/2005 | 110,000 | 1.25 | 69,000 | 110,000 | Competitive |
| Las Vegas Field Office | Lizette Harvey | 12/07/2005 | 117,500 | 1.25 | 69,000 | 117,500 | Competitive |
| Las Vegas Field Office | Muscle Investments, LLC | 01/05/2006 | 105,000 | 1.25 | 69,000 | 105,000 | Competitive |
| Las Vegas Field Office | The Crescent Group, LLC | 12/09/2005 | 110,000 | 1.25 | 69,000 | 110,000 | Competitive |

**Appendix II. Continued**

| State and office | Customer | Transaction date | Amount collected | Acreage | Fair market value | Sale price | Disposal method |
|---|---|---|---|---|---|---|---|
| Las Vegas Field Office | Mary Borges | 12/09/2005 | 115,000 | 1.25 | 69,000 | 115,000 | Competitive |
| Las Vegas Field Office | Sean Afshar | 12/12/2005 | 130,000 | 1.25 | 69,000 | 130,000 | Competitive |
| Las Vegas Field Office | Edward Van Sloten | 06/16/2005 | 160,000 | 1.25 | 69,000 | 160,000 | Competitive |
| Las Vegas Field Office | Jeffry Yelland | 06/16/2005 | 170,000 | 1.25 | 69,000 | 170,000 | Competitive |
| Las Vegas Field Office | Varinder Singh | 12/12/2005 | 150,000 | 1.25 | 69,000 | 150,000 | Competitive |
| Las Vegas Field Office | William Berdie Jr. | 12/05/2005 | 125,000 | 1.25 | 69,000 | 125,000 | Competitive |
| Las Vegas Field Office | Sean Afshar | 12/12/2005 | 137,500 | 1.25 | 69,000 | 137,500 | Competitive |
| Las Vegas Field Office | John Hadley | 12/06/2005 | 100,000 | 1.25 | 69,000 | 100,000 | Competitive |
| Las Vegas Field Office | Ermila Picking | 06/16/2005 | 26,000 | 1.25 | 69,000 | 26,000 | Competitive |
| Las Vegas Field Office | Sean Afshar | 12/12/2005 | 130,000 | 1.25 | 69,000 | 130,000 | Competitive |
| Las Vegas Field Office | Louise Myers | 12/07/2005 | 130,000 | 1.25 | 69,000 | 130,000 | Competitive |
| Las Vegas Field Office | Varinder Singh | 12/12/2005 | 165,000 | 1.25 | 69,000 | 165,000 | Competitive |
| Las Vegas Field Office | Sean Afshar | 12/12/2005 | 200,000 | 1.25 | 69,000 | 200,000 | Competitive |
| Las Vegas Field Office | Sean Afshar | 12/12/2005 | 167,500 | 1.25 | 69,000 | 167,500 | Competitive |
| Las Vegas Field Office | Sean Afshar | 12/12/2005 | 172,500 | 1.25 | 69,000 | 172,500 | Competitive |
| Las Vegas Field Office | Sean Afshar | 12/12/2005 | 172,500 | 1.25 | 69,000 | 172,500 | Competitive |
| Las Vegas Field Office | James Hunter | 12/08/2005 | 135,000 | 1.25 | 69,000 | 135,000 | Competitive |
| Las Vegas Field Office | James Hunter | 12/08/2005 | 140,000 | 1.25 | 69,000 | 140,000 | Competitive |
| Las Vegas Field Office | Shane and Karen Lapier | 10/27/2005 | 125,000 | 1.25 | 69,000 | 125,000 | Competitive |
| Las Vegas Field Office | Joanne Phillips | 06/16/2005 | 130,000 | 1.25 | 69,000 | 130,000 | Competitive |
| Las Vegas Field Office | Sean Afshar | 12/12/2005 | 187,500 | 1.25 | 69,000 | 187,500 | Competitive |
| Las Vegas Field Office | Sean Afshar | 12/12/2005 | 202,500 | 1.25 | 69,000 | 202,500 | Competitive |

**Appendix II. Continued**

| State and office | Customer | Transaction date | Amount collected | Acreage | Fair market value | Sale price | Disposal method |
|---|---|---|---|---|---|---|---|
| Las Vegas Field Office | Sean Afshar | 12/12/2005 | 194,000 | 1.25 | 69,000 | 194,000 | Competitive |
| Las Vegas Field Office | Sean Afshar | 12/12/2005 | 195,000 | 1.25 | 69,000 | 195,000 | Competitive |
| Las Vegas Field Office | Ominet Laughlin, LLC | 05/12/2006 | 8,000,000 | 80 | 7,040,000 | 8,000,000 | Competitive |
| Las Vegas Field Office | Laughlin Properties, LLC | 12/08/2005 | 2,275,000 | 20 | 1,860,000 | 2,275,000 | Competitive |
| Las Vegas Field Office | Terraspec Development, LLC | 12/09/2005 | 850,000 | 5 | 540,000 | 850,000 | Competitive |
| Las Vegas Field Office | DJL Enterprises, LLC | 12/06/2005 | 5,150,000 | 50 | 5,150,000 | 5,150,000 | Competitive |
| Las Vegas Field Office | DJL Enterprises, LLC | 12/06/2005 | 7,275,000 | 40 | 4,800,000 | 7,275,000 | Competitive |
| Las Vegas Field Office | D.J. Laughlin | 12/06/2005 | 4,140,000 | 20 | 2,280,000 | 4,140,000 | Competitive |
| Las Vegas Field Office | DJL Enterprises, LLC | 12/06/2005 | 3,325,000 | 20 | 2,160,000 | 3,325,000 | Competitive |
| Las Vegas Field Office | Peter Horne dba Halo Realty Investments | 08/29/2005 | 202,500 | 2.9 | 128,000 | 202,500 | Competitive |
| Las Vegas Field Office | Hardy Properties, LLC | 12/08/2005 | 170,000 | 2.91 | 127,000 | 170,000 | Competitive |
| Las Vegas Field Office | Lily Ripley | 12/12/2005 | 95,000 | 2.9 | 93,000 | 95,000 | Competitive |
| Las Vegas Field Office | Skyline West Investments | 12/12/2005 | 187,500 | 2.9 | 142,000 | 187,500 | Competitive |
| Las Vegas Field Office | Hardy Properties, LLC | 12/08/2005 | 182,500 | 2.92 | 131,000 | 182,500 | Competitive |
| Las Vegas Field Office | Brooke Ann Mrofcza | 12/08/2005 | 153,000 | 2.93 | 117,000 | 153,000 | Competitive |

**Appendix II. Continued**

| State and office | Customer | Transaction date | Amount collected | Acreage | Fair market value | Sale price | Disposal method |
|---|---|---|---|---|---|---|---|
| Las Vegas Field Office | Brooke Ann Mrofcza | 12/08/2005 | 161,000 | 2.92 | 131,000 | 161,000 | Competitive |
| Las Vegas Field Office | Silver State Schools Credit Union | 11/08/2005 | 92,000 | 2.92 | 88,000 | 92,000 | Competitive |
| Las Vegas Field Office | Foundation Capital Preservation Trust | 12/22/2005 | 160,000 | 2.91 | 131,000 | 160,000 | Competitive |
| Las Vegas Field Office | Hardy Properties, LLC | 12/08/2005 | 167,000 | 2.92 | 131,000 | 167,000 | Competitive |
| Las Vegas Field Office | Denise Ware | 09/19/2005 | 220,000 | 2.91 | 131,000 | 220,000 | Competitive |
| Las Vegas Field Office | Hardy Properties, LLC | 12/08/2005 | 200,000 | 2.91 | 160,000 | 200,000 | Competitive |
| Las Vegas Field Office | Hardy Properties, LLC | 12/08/2005 | 150,000 | 2.9 | 90,000 | 150,000 | Competitive |
| Las Vegas Field Office | Hardy Properties, LLC | 12/08/2005 | 170,000 | 2.9 | 131,000 | 170,000 | Competitive |
| Las Vegas Field Office | Foundation Capital Preservation Trust | 12/22/2005 | 161,000 | 2.9 | 64,000 | 161,000 | Competitive |
| Las Vegas Field Office | Melvin Henry Cavanaugh Jr. | 12/12/2005 | 112,000 | 2.89 | 52,000 | 112,000 | Competitive |
| Tonopah Field Station | The Botner 1992 Family Trust | 07/16/2004 | 5,500 | 1.35 | 5,500 | 5,500 | Direct |
| Tonopah Field Station | Fred and Florita McMillen III | 12/18/2003 | 42,000 | 7.5 | 42,000 | 42,000 | Direct |
| Tonopah Field Station | Nye County Commissioners | 06/26/2003 | 180,000 | 11.71 | 180,000 | 180,000 | Direct |
| Tonopah Field Station | James Key | 07/15/2003 | 16,000 | 2.5 | 16,000 | 16,000 | Direct |
| Tonopah Field Station | John Maurer | 03/07/2003 | 16,000 | 80 | 16,000 | 16,000 | Direct |

**Appendix II. Continued**

| State and office | Customer | Transaction date | Amount collected | Acreage | Fair market value | Sale price | Disposal method |
|---|---|---|---|---|---|---|---|
| Tonopah Field Station | Rockview Dairies, Inc. | 05/04/2004 | 480,000 | 320 | 480,000 | 480,000 | Modified competitive |
| Tonopah Field Station | Ponderosa Dairy | 04/22/2004 | 144,000 | 120 | 144,000 | 144,000 | Direct |
| Winnemucca Field Office | Verna Wallace | 07/30/2003 | 11,000 | 40 | 11,000 | 11,000 | Direct |
| Winnemucca Field Office | Pershing County, Nevada | 07/27/2000 | 44,000 | 350 | 44,000 | 44,000 | Direct |
| Winnemucca Field Office | RDD, Inc. | 11/24/2000 | 71,500 | 953.56 | 71,500 | 71,500 | Direct |
| Winnemucca Field Office | W. Krys Palulis | 04/29/2003 | 31,200 | 80 | 26,000 | 28,080 | Competitive |
| Winnemucca Field Office | Vince Wavra | 06/14/2005 | 28,000 | 160 | 28,000 | 28,000 | Competitive |
| Winnemucca Field Office | Home Ranch, LLC | 06/14/2005 | 56,000 | 319.95 | 28,000 | 56,000 | Competitive |
| Winnemucca Field Office | Vince Wavra | 06/14/2005 | 29,047 | 160.48 | 28,084 | 29,047 | Competitive |
| Winnemucca Field Office | Home Ranch, LLC | 06/14/2005 | 42,000 | 240 | 42,000 | 42,000 | Competitive |
| **New Mexico** | | | | | | | |
| Carlsbad Field Office | Ray Westall | 07/31/2001 | 20,000 | 5 | 20,000 | 20,000 | Direct |
| Farmington Field Office | Clifford Martinez | 12/18/2003 | 25,700 | 4.6 | 25,700 | 25,700 | Direct |
| Farmington Field Office | Charles and Joan Eavenson | 12/19/2002 | 10,600 | 0.52 | 10,600 | 10,600 | Direct |
| Farmington Field Office | Tonie Martinez | 10/18/2004 | 18,000 | 3.6 | 18,000 | 18,000 | Direct |
| Las Cruces District Office | Our Lady's Youth Center | 02/26/2002 | 230,400 | 320 | 288,000 | 256,000 | Direct |
| Las Cruces District Office | Mesa Farmers Coop | 04/19/2006 | 2,070,000 | 396.34 | 879,000 | 2,070,000 | Competitive |
| Las Cruces District Office | Philippou, LLC | 01/20/2005 | 1,600,000 | 39.47 | 850,000 | 1,600,000 | Competitive |

## Appendix II. Continued

| State and office | Customer | Transaction date | Amount collected | Acreage | Fair market value | Sale price | Disposal method |
|---|---|---|---|---|---|---|---|
| Pecos District Office | Joe Cox | 02/16/2006 | 2,000 | 5 | 2,000 | 2,000 | Direct |
| Taos Field Office | Robert Montoya | 02/26/2002 | 15,000 | 0.38 | 15,000 | 15,000 | Direct |
| Taos Field Office | Manuel Vigil | 02/26/2002 | 500 | 0.13 | 500 | 500 | Direct |
| Taos Field Office | Robert Anaya | 04/15/2003 | 20,840 | 1.33 | 29,600 | 20,840 | Direct |
| Taos Field Office | Joseph Chipman | 09/06/2006 | 28,000 | 0.5 | 28,000 | 28,000 | Direct |
| Taos Field Office | Heirs of Benerito Ortega | 03/07/2006 | 8,470 | 1.2 | 8,470 | 8,470 | Direct |
| Taos Field Office | Lafayette Rodriguez | 04/19/2006 | 3,290 | 0.4 | 3,290 | 3,290 | Direct |
| **Oregon** | | | | | | | |
| Burns District Office | American Exchange Services, Inc. | 02/26/2002 | 36,000 | 360 | 36,000 | 36,000 | Competitive |
| Burns District Office | Delmer Clemens | 02/26/2002 | 20,888 | 399.19 | 26,000 | 26,110 | Competitive |
| Burns District Office | Delmer and Teresa Clemens | 02/26/2002 | 5,200 | 80.63 | 5,200 | 5,200 | Competitive |
| Burns District Office | Jon Woodworth | 02/26/2002 | 12,560 | 157 | 15,700 | 15,700 | Competitive |
| Burns District Office | John and Judy Ahmann | 02/26/2002 | 6,451 | 80 | 8,000 | 8,101 | Competitive |
| Burns District Office | Fort Harney Ranch | 02/26/2002 | 14,400 | 246.88 | 16,000 | 18,000 | Competitive |
| Burns District Office | Louis Borelli III | 02/26/2002 | 15,000 | 80 | 8,000 | 15,000 | Competitive |
| Burns District Office | Marcia Eggleston | 02/26/2002 | 3,800 | 40 | 3,600 | 3,800 | Modified competitive |
| Burns District Office | Marcia Eggleston | 02/26/2002 | 4,000 | 40 | 3,600 | 4,000 | Modified competitive |
| Burns District Office | Richard Boatwright Jr. | 02/26/2002 | 36,000 | 200 | 20,000 | 36,000 | Competitive |
| Burns District Office | Thomas and Barbara Howard | 02/26/2002 | 29,000 | 320 | 28,800 | 29,000 | Modified competitive |

**Appendix II. Continued**

| State and office | Customer | Transaction date | Amount collected | Acreage | Fair market value | Sale price | Disposal method |
|---|---|---|---|---|---|---|---|
| Burns District Office | Thomas and Barbara Howard | 02/26/2002 | 4,000 | 40 | 3,600 | 4,000 | Competitive |
| Burns District Office | Bill Wilber | 02/26/2002 | 8,000 | 80 | 8,000 | 8,000 | Competitive |
| Burns District Office | Ray Drayton | 09/10/2003 | 8,400 | 79.79 | 5,600 | 8,400 | Competitive |
| Burns District Office | George Davis | 02/26/2002 | 1,860 | 79.79 | 7,200 | a | Competitive |
| Burns District Office | Bell A. Grazing | 02/26/2002 | 14,400 | 160 | 14,400 | 14,400 | Modified competitive |
| Burns District Office | Bell A. Grazing | 02/26/2002 | 10,800 | 120 | 10,800 | 10,800 | Modified competitive |
| Burns District Office | Bell A. Grazing | 02/26/2002 | 3,600 | 40 | 3,600 | 3,600 | Modified competitive |
| Burns District Office | Martin and Andrea Davies | 02/26/2002 | 7,200 | 80 | 7,200 | 7,200 | Modified competitive |
| Burns District Office | Maurice and Norma Davies | 02/26/2002 | 8,000 | 80 | 8,000 | 8,000 | Competitive |
| Burns District Office | Jon Cicin | 02/26/2002 | 8,250 | 80 | 7,200 | 8,250 | Competitive |
| Burns District Office | Van Grazing Cooperative, Inc. | 10/28/2002 | 24,000 | 120 | 24,000 | 24,000 | Modified competitive |
| Burns District Office | Van Grazing Cooperative, Inc. | 07/01/2002 | 8,000 | 40 | 8,000 | 8,000 | Modified competitive |
| Burns District Office | David and Rachel Boyd | 09/11/2003 | 40,000 | 320 | 40,000 | 40,000 | Competitive |
| Burns District Office | Lynn Deguire | 02/24/2006 | 21,202 | 145.56 | 21,100 | 21,202 | Competitive |
| Burns District Office | Lynn Deguire | 12/08/2005 | 3,602 | 40 | 3,600 | 3,602 | Competitive |
| Burns District Office | Gary Marshall | 02/06/2007 | 3,710 | 40 | 3,400 | 3,710 | Competitive |
| Burns District Office | Dylan Decelis | 02/09/2005 | 650 | 40 | 3,200 | a | Competitive |

## Appendix II. Continued

| State and office | Customer | Transaction date | Amount collected | Acreage | Fair market value | Sale price | Disposal method |
|---|---|---|---|---|---|---|---|
| Burns District Office | Ray Drayton | 01/02/2004 | 11,700 | 160 | 11,200 | 11,700 | Competitive |
| Burns District Office | Carolyn and David Mooers | 11/12/2003 | 87,000 | 160 | 24,800 | 87,000 | Competitive |
| Burns District Office | Vera Hotchkiss | 08/13/2003 | 15,900 | 109.42 | 15,900 | 15,900 | Competitive |
| Burns District Office | DJ Miller Ranches | 08/14/2003 | 15,258 | 80 | 12,800 | 15,258 | Competitive |
| Burns District Office | Jerry Temple | 02/24/2006 | 3,000 | 40 | 3,000 | 3,000 | Modified competitive |
| Burns District Office | Jerry Temple | 01/23/2004 | 3,050 | 40 | 3,000 | 3,050 | Modified competitive |
| Burns District Office | Tom and Diane Grant Jr. | 09/10/2003 | 3,000 | 40 | 2,600 | 3,000 | Competitive |
| Burns District Office | Tyler Brothers | 12/11/2003 | 30,000 | 160 | 16,600 | 30,000 | Competitive |
| Burns District Office | Tyler Brothers | 12/11/2003 | 15,000 | 160 | 11,600 | 15,000 | Modified competitive |
| Burns District Office | Tyler Brothers | 12/11/2003 | 3,050 | 40.62 | 3,050 | 3,050 | Modified competitive |
| Burns District Office | Tyler Brothers | 02/24/2006 | 5,800 | 80 | 5,800 | 5,800 | Modified competitive |
| Burns District Office | Joseph and Lois Eckley | 01/13/2004 | 8,500 | 80 | 6,000 | 8,500 | Competitive |
| Burns District Office | Ray Drayton | 09/10/2003 | 12,000 | 120 | 8,400 | 12,000 | Competitive |
| Burns District Office | Ray Drayton | 01/02/2004 | 3,310 | 40 | 3,000 | 3,310 | Competitive |
| Burns District Office | Joseph and Lois Eckley | 08/14/2003 | 3,150 | 39.18 | 3,150 | 3,150 | Competitive |
| Burns District Office | Zack Sword | 02/12/2004 | 8,400 | 120 | 8,400 | 8,400 | Modified competitive |
| Burns District Office | Shirley Thompson | 02/05/2004 | 3,200 | 40 | 3,200 | 3,200 | Modified competitive |
| Burns District Office | Zack Sword | 02/12/2004 | 3,200 | 40 | 3,200 | 3,200 | Modified competitive |
| Burns District Office | John Ahmann | 12/14/2006 | 3,200 | 80 | 16,000 | 16,000 | Modified competitive |
| Burns District Office | William Dunbar | 01/22/2007 | 22,800 | 119.53 | 22,700 | 22,800 | Competitive |
| Burns District Office | Gerard Labrecque | 12/13/2006 | 1,000 | 0.51 | 200 | 1,000 | Modified competitive |

## Appendix II. Continued

| State and office | Customer | Transaction date | Amount collected | Acreage | Fair market value | Sale price | Disposal method |
|---|---|---|---|---|---|---|---|
| Burns District Office | Ray Drayton | 03/08/2007 | 48,500 | 160 | 21,600 | 48,500 | Competitive |
| Burns District Office | Daniel and Denise Kryger | 12/13/2006 | 18,400 | 240 | 58,100 | 92,000 | Competitive |
| Burns District Office | Terry and Donald Cutsforth/Wagner | 05/23/2007 | 32,000 | 80 | 9,100 | 32,000 | Competitive |
| Burns District Office | G. Edward Barnes | 05/17/2007 | 16,100 | 79.81 | 9,200 | 16,100 | Competitive |
| Burns District Office | Terry and Donald Cutsforth/Wagner | 05/23/2007 | 46,000 | 119.76 | 19,100 | 46,000 | Competitive |
| Burns District Office | Tracy Hill | 01/30/2007 | 16,800 | 160 | 16,800 | 16,800 | Competitive |
| Burns District Office | Ray Drayton | 12/13/2006 | 12,780 | 200 | 19,600 | 63,900 | Competitive |
| Burns District Office | Morris Family Trust | 01/18/2007 | 8,000 | 80 | 8,000 | 8,000 | Competitive |
| Burns District Office | Heather Harris | 05/24/2007 | 8,000 | 80 | 8,000 | 8,000 | Competitive |
| Burns District Office | Joseph Stefanowitz | 03/07/2007 | 56,100 | 80 | 9,800 | 56,100 | Competitive |
| Burns District Office | Bell A. Grazing | 05/14/2007 | 6,800 | 80 | 6,800 | 6,800 | Modified competitive |
| Burns District Office | Bell A. Grazing | 05/14/2007 | 3,900 | 40 | 3,900 | 3,900 | Modified competitive |
| Burns District Office | Shirley Thompson | 12/14/2006 | 1,990 | 120 | 9,700 | 9,700 | Modified competitive |
| Burns District Office | Shirley Thompson | 12/14/2006 | 690 | 40 | 3,200 | 3,200 | Modified competitive |
| Burns District Office | Shirley Thompson | 12/14/2006 | 690 | 40 | 3,200 | 3,200 | Modified competitive |
| Burns District Office | James Drayton | 03/30/2007 | 73,200 | 440 | 48,400 | 73,200 | Competitive |
| Burns District Office | Peter Barry | 12/14/2006 | 9,400 | 240 | 28,800 | 47,000 | Competitive |
| Burns District Office | Stanley and Gale Kazebee/Nelson | 12/14/2006 | 30,200 | 160 | 18,400 | 151,000 | Competitive |

**Appendix II. Continued**

| State and office | Customer | Transaction date | Amount collected | Acreage | Fair market value | Sale price | Disposal method |
|---|---|---|---|---|---|---|---|
| Burns District Office | Rattlesnake Creek Land and Cattle Company, LLC | 12/14/2006 | 6,640 | 185.81 | 33,100 | 33,200 | Modified competitive |
| Burns District Office | Stanley Kull | 12/18/2006 | 7,200 | 40 | 7,200 | 7,200 | Modified competitive |
| Eugene District Office | Harold Leahy and Kieran, Attorneys At Law | 10/10/2002 | 2,350 | 3.74 | 22,700 | 23,500 | Modified competitive |
| Eugene District Office | Western Pioneer Title Company | 03/05/2002 | 1,500 | 1.72 | 1,500 | 1,500 | Direct |
| Eugene District Office | James Bean | 08/21/2006 | 10,000 | 0.45 | 10,000 | 10,000 | Direct |
| Eugene District Office | Tony and Sonya Bratton | 12/04/2003 | 1,100 | 0.16 | 1,100 | 1,100 | Direct |
| Lakeview District Office | Donald Rajnus | 02/26/2002 | 10,000 | 80 | 10,000 | 10,000 | Direct |
| Lakeview District Office | Kennedy Land Co, LLC | 12/09/2002 | 10,900 | 119.76 | 10,900 | 10,900 | Direct |
| Lakeview District Office | Meadow Lake, Inc. | 02/26/2002 | 18,200 | 120.12 | 18,200 | 18,200 | Direct |
| Lakeview District Office | Virginia Rajnus | 02/26/2002 | 5,600 | 80 | 5,600 | 5,600 | Direct |
| Lakeview District Office | Alston Bruner | 12/02/2004 | 3,600 | 40 | 3,600 | 3,600 | Modified competitive |
| Lakeview District Office | Alan Withers | 05/24/2006 | 4,000 | 40.24 | 4,000 | 4,000 | Direct |
| Prineville District Office | Mont West | 02/26/2002 | 1,800 | 0.43 | 1,800 | 1,800 | Direct |
| Roseburg District Office | Douglas County | 08/17/2004 | 7,254 | 0.08 | 7,254 | 7,254 | Direct |
| Vale District Office | George and Joanne Voile | 02/10/2005 | 3,300 | 11.25 | 3,300 | 3,300 | Direct |

# Appendix II. Continued

| State and office | Customer | Transaction date | Amount collected | Acreage | Fair market value | Sale price | Disposal method |
|---|---|---|---|---|---|---|---|
| **Utah** | | | | | | | |
| St. George Field Office | City of St. George | 02/26/2002 | 177,000 | 26.18 | 177,000 | 177,000 | Direct |
| **Wyoming** | | | | | | | |
| Lander Field Office | State of Wyoming | 03/21/2003 | 1,500 | 10.27 | 1,500 | 1,500 | Direct |
| Newcastle Field Office | Dennis Drayton | 02/26/2002 | 7,200 | 80 | 7,200 | 7,200 | Competitive |
| Newcastle Field Office | John Hiller | 02/26/2002 | 3,650 | 40 | 3,650 | 3,650 | Competitive |
| Newcastle Field Office | W. O'Kief | 02/26/2002 | 3,600 | 40 | 3,600 | 3,600 | Competitive |
| Newcastle Field Office | Thomas Randall | 02/26/2002 | 3,611 | 40 | 3,611 | 3,611 | Competitive |
| Newcastle Field Office | George Paul | 01/17/2003 | 28,200 | 39.56 | 28,200 | 28,200 | Modified competitive |
| Newcastle Field Office | Richard and Beth Schuetz | 11/12/2002 | 36,800 | 42.44 | 36,800 | 36,800 | Modified competitive |
| Pinedale Field Office | William Mayo | 02/26/2002 | 16,600 | 40 | 16,600 | 16,600 | Direct |
| Rawlins Field Office | Baggs Solid Waste Disposal District | 02/26/2002 | 9,600 | 120 | 3,600 | 3,600 | Direct |
| Rock Springs Field Office | David J. Palmer | 02/26/2002 | 2,200 | 0.06 | 2,200 | 2,200 | Direct |
| Rock Springs Field Office | PacifiCorp | 05/03/2004 | 722,500 | 722.5 | 722,500 | 722,500 | Direct |
| Worland Field Office | Mary A. Clay Revocable Trust | 03/20/2006 | 1,924 | 3.75 | 1,924 | 1,924 | Direct |
| Worland Field Office | Robert Gilmore Griffin | 08/19/2002 | 2,700 | 30 | 2,700 | 2,700 | Direct |

Source: GAO analysis of information from BLM Division of Business Services and state offices.

[a]Information not provided.

*Appendix III*

# DETAILED INFORMATION ON PLANNED FLTFA LAND SALES THROUGH 2010, AS REPORTED BY BLM STATE OFFICES

## Appendix III.

| State/Field Office | Acreage | Fair market value | Sale method | Disposal reason |
|---|---|---|---|---|
| **Arizona** | | | | |
| Arizona Strip Field Office | 118.8 | $[a] | Competitive | Management inefficiency |
| Hassayampa Field Office | 1,030 | [a] | Competitive | Community development |
| Lower Sonoran Field Office | 17 | [a] | Direct | Trespass resolution |
| Lower Sonoran Field Office | 55.5 | 110,900 | Direct | Management inefficiency |
| Tucson Field Office | 69 | [a] | Modified competitive | Trespass resolution |
| Yuma Field Office | 1,000 | [a] | Competitive | Community development |
| Yuma Field Office | 300 | [a] | Competitive | Management inefficiency |
| Yuma Field Office | 50 | [a] | Competitive | Community development |
| **California** | | | | |
| Bishop Field Office | 0.25 | [a] | Direct | Difficult and uneconomic to manage, inadvertent trespass |
| Folsom Field Office | 11 | [a] | Direct | Difficult and uneconomic to manage, inadvertent trespass |
| Redding Field Office | 0.07 | [a] | Direct | Difficult and uneconomic to manage fragment, inadvertent trespass |
| Redding Field Office | 80 | [a] | Competitive | Not needed for federal purposes, difficult and uneconomic to manage |
| Redding Field Office | 160 | [a] | Direct | Not needed for federal purposes, buffer to landfill |

## Appendix III. Continued

| State/Field Office | Acreage | Fair market value | Sale method | Disposal reason |
|---|---|---|---|---|
| **Colorado** | | | | |
| Gunnison Field Office | 0.76 | 7,000 | Direct | Inadvertent occupancy of public land |
| Royal Gorge Field Office | 0.09 | a | Direct | To resolve historic, unauthorized residential use |
| Royal Gorge Field Office | 0.40 | a | Direct | To resolve historic, unauthorized residential use |
| Royal Gorge Field Office | 0.53 | a | Direct | To resolve historic, unauthorized residential use |
| Royal Gorge Field Office | 2.99 | a | Direct | To resolve historic, unauthorized residential use |
| Royal Gorge Field Office | 1.61 | a | Direct | To resolve historic, unauthorized residential use |
| Royal Gorge Field Office | 0.39 | a | Direct | To resolve historic, unauthorized residential use |
| Royal Gorge Field Office | 0.42 | a | Direct | To resolve historic, unauthorized residential use |
| Royal Gorge Field Office | 0.17 | a | Direct | To resolve historic, unauthorized residential use |
| Royal Gorge Field Office | 0.44 | a | Direct | To resolve historic, unauthorized residential use |

## Appendix III. Continued

| State/Field Office | Acreage | Fair market value | Sale method | Disposal reason |
|---|---|---|---|---|
| Royal Gorge Field Office | 0.27 | a | Direct | To resolve historic, unauthorized residential use |
| Royal Gorge Field Office | 2.09 | a | Direct | To resolve historic, unauthorized residential use |
| Royal Gorge Field Office | 7 | a | Competitive | To resolve historic, unauthorized residential use |
| San Juan Public Lands Center | 36.92 | a | Undetermined | Meets resource management plan criteria for disposal |
| San Juan Public Lands Center | 40 | a | Direct | Land needed by county |
| San Juan Public Lands Center | 2.11 | a | Direct | Strip 40 ft. X 2560 ft. landowner needs |
| San Juan Public Lands Center | 40 | 248,000 | Direct | Meets resource management plan criteria for disposal |
| **Idaho** | | | | |
| Challis Field Office | 80 | a | Direct | Existing shooting range |
| Challis Field Office | 100 | a | Direct | Waste transfer site |
| Challis Field Office | 11.34 | a | Direct | Existing waste disposal site |
| Challis Field Office | 127 | a | Direct | Existing waste disposal site |
| Challis Field Office | 302 | a | Direct | Difficult and uneconomic to manage |
| Challis Field Office | 104 | 172,000 | Modified competitive | Difficult and uneconomic to manage |
| Coeur d'Alene Field Office | 2.97 | 50,000 | Direct | Color of Title Act |
| Coeur d'Alene Field Office | 9.4 | 159,800 | Direct | Color of Title Act |

**Appendix III. Continued**

| State/Field Office | Acreage | Fair market value | Sale method | Disposal reason |
|---|---|---|---|---|
| Coeur d'Alene Field Office | 1 | a | Direct | Trespass |
| Coeur d'Alene Field Office | 5.07 | a | Direct | Trespass |
| Four Rivers Field Office | 2,995.54 | 721,361 | Competitive | Isolated |
| Four Rivers Field Office | 80 | 1,600,000 | Competitive | Isolated |
| Shoshone Field Office | 120 | a | Direct | a |
| Shoshone Field Office | 1.62 | a | Direct | a |
| Shoshone Field Office | 40 | a | Direct | a |
| Shoshone Field Office | 262.21 | a | Direct | a |
| **Nevada** | | | | |
| Battle Mountain Field Office | 0.5 | 12,500 | Direct | a |
| Battle Mountain Field Office | 569.34 | 115,000 | Modified competitive | a |
| Battle Mountain Field Office | 2.65 | 20,000 | Direct | a |
| Battle Mountain Field Office | 878.34 | 439,000 | Competitive | a |
| Carson City Field Office | 56.25 | 1,450,000 | Competitive | a |
| Carson City Field Office | 370 | 165,000 | Direct | a |
| Carson City Field Office | 628.2 | 1,300,000 | Competitive | a |
| Carson City Field Office | 240 | 6,000,000 | Competitive | a |
| Elko Field Office | 640 | 130,000 | Competitive | a |
| Elko Field Office | 640 | 120,000 | Competitive | a |
| Elko Field Office | 640 | 125,000 | Competitive | a |
| Elko Field Office | 640 | 115,000 | Competitive | a |

## Appendix III. Continued

| State/Field Office | Acreage | Fair market value | Sale method | Disposal reason |
|---|---|---|---|---|
| Elko Field Office | 640 | 110,000 | Competitive | a |
| Elko Field Office | 360 | 75,000 | Competitive | a |
| Elko Field Office | 1,440 | 200,000 | Competitive | a |
| Elko Field Office | 2,560 | 300,000 | Competitive | a |
| Elko Field Office | 80 | 50,000 | Competitive | a |
| Ely Field Office | 217 | 217,000 | Direct | a |
| Ely Field Office | 159 | 160,000 | Competitive | a |
| Las Vegas Field Office | 292.46 | 965,118 | Direct | a |
| Winnemucca Field Office | 382.5 | 396,000 | Direct | a |
| Winnemucca Field Office | 1,440 | 144,000 | Competitive | a |
| Winnemucca Field Office | 440 | 44,000 | Competitive | a |
| Winnemucca Field Office | 1,214 | 212,000 | Competitive | a |
| Winnemucca Field Office | 40 | 120,000 | Direct | a |
| **New Mexico** | | | | |
| Farmington Field Office | 80 | 2,000,000 | Direct | Competitive public interest, expansion of community and economic development |
| Las Cruces District Office | 166.59 | a | Competitive | Competitive public interest, expansion of community and economic development |
| Las Cruces District Office | 180 | a | Competitive | Competitive public interest, expansion of community and economic development |

## Appendix III. Continued

| State/Field Office | Acreage | Fair market value | Sale method | Disposal reason |
|---|---|---|---|---|
| Las Cruces District Office | 139.77 | [a] | Competitive | Competitive public interest, expansion of community and economic development |
| Las Cruces District Office | 140.39 | [a] | Competitive | Competitive public interest, expansion of community and economic development |
| Las Cruces District Office | 400 | 3,500,000 | Competitive | Competitive public interest, expansion of community and economic development |
| Las Cruces District Office | 160 | [a] | Competitive | Competitive public interest, expansion of community and economic development |
| Rio Puerco Field Office | 0.48 | [a] | Direct | To resolve an unintentional, unauthorized occupancy |
| Taos Field Office | 6 | [a] | Competitive | Competitive public interest, expansion of community and economic development |
| **Utah** | | | | |
| Kanab Field Office | 120 | 170,000 | Modified competitive | Isolated parcel |
| Moab Field Office | 50 | 50,000 | Direct | To resolve unauthorized use (agricultural) |
| Richfield Field Office | 4.82 | [a] | Direct | To resolve unauthorized use (occupancy) |
| St. George Field Office | 145 | 14,000,000 | Competitive | [a] |
| Vernal Field Office | 62.5 | 1,000 | Direct | Accommodate use on adjoining lands (sewage treatment ponds) |
| Vernal Field Office | 30 | 13,450 | Direct | Resolve unauthorized use (agricultural) |

## Appendix III. Continued

| State/Field Office | Acreage | Fair market value | Sale method | Disposal reason |
|---|---|---|---|---|
| **Wyoming** | | | | |
| Casper Field Office | 20 | a | Direct | Community expansion |
| Kemmerer Field Office | 120 | a | Direct | Landfill expansion |
| Newcastle Field Office | 1.34 | a | Direct | Resolution of trespass |
| Newcastle Field Office | 2.27 | a | Competitive | Community expansion |
| Rawlins Field Office | 640 | a | Modified competitive | Community expansion |
| Rawlins Field Office | 40 | a | Modified competitive | Community expansion |
| Rock Springs Field Office | 40 | a | Direct | Resolution of an occupancy trespass |
| Rock Springs Field Office | 820 | a | Direct | Plant expansion |
| Rock Springs Field Office | 39.1 | a | Undetermined | Resolution of an occupancy trespass |
| Rock Springs Field Office | 157.24 | a | Competitive | Industrial expansion |

Source: GAO analysis of information from BLM's state offices.

[a]Information not provided.

# REFERENCES

[1]  U.S. Department of the Interior and U.S. Department of Agriculture, *National Land Acquisition Plan* (Washington, D.C., February 2005). The agencies estimated the following acres of inholdings: National Park Service—6.5 million acres; U.S. Fish and Wildlife Service—17.3 million acres; Forest Service—about 40 million acres; and BLM—over 7 million acres within national monuments and national conservation areas and several million more in other areas.

[2]  Pub. L. 106-248 (2000) (codified as 43 U.S.C. § 2301 et seq).

[3]  Land exchanges sometimes result in the exchange of unequally valued land. In such cases, the nonfederal entity generally pays the difference to the federal agency in the form of a cash equalization payment. Proceeds from such transactions involving FLTFA-eligible lands are deposited into the FLTFA account.

[4]  The authority is provided in the Federal Land Policy and Management Act (FLPMA) of 1976 (Pub. L. 94-579) (1976) (codified at 43 U.S.C. § 1701 et seq.). FLPMA defines the 11 contiguous western states as Arizona, California, Colorado, Idaho, Montana, Nevada, New Mexico, Oregon, Utah, Washington, and Wyoming (43 U.S.C. 1702, § 103 (o)). BLM's Alaska State Office is not currently participating in FLTFA because of its priority to settle Alaska Native land claims.

[5]  BLM land use plans may also be called "resource management plans" and "management framework plans." We will refer to them as land use plans, the term used in FLTFA.

[6]  This new account is referred to in the act as the Federal Land Disposal Account. Before FLTFA, revenue from these transactions were typically deposited into the U.S. Treasury's general account.

[7]     According to the act (43 U.S.C. 2302), the term "exceptional resource" means a resource of scientific, natural, historic, cultural, or recreational value that has been documented by a federal, state, or local governmental authority, and for which there is a compelling need for conservation and protection under the jurisdiction of a Federal agency in order to maintain the resource for the benefit of the public.

[8]     For this book, "field offices" refers to BLM's 26 district offices and 118 field offices.

[9]     Effective October 1, 2007, a reorganization of the BLM centers in Denver merged the National Business Center, National Human Resources Management Center, National Information Resources Management Center, and National Science and Technology Center into a single unit called the National Operations Center. The National Business Center is now known as the Division of Business Services.

[10]    This refers to Title I of Pub. L. 106-248, which provides for the federal acquisition of the Baca Ranch in New Mexico. FLTFA is Title II of this act.

[11]    The Land and Water Conservation Fund Act of 1965 established the Land and Water Conservation Fund (Pub. L. 88-578). Among other sources of land acquisition funding is the Migratory Bird Fund used exclusively by the Fish and Wildlife Service.

[12]    The LWCF is a trust fund that accumulates revenue from federal outdoor recreation user fees, the federal motorboat fuel tax, and surplus property sales. To supplement these sources to reach its annual authorized level of $900 million, the fund accumulates revenue from oil and gas leases on the Outer Continental Shelf.

[13]    By comparison, in fiscal year 2006, Congress appropriated $41.8 million, or about 35 percent, to the Forest Service; $34.4 million, or about 29 percent, to the Park Service; $28.0 million, or about 23 percent, to the Fish and Wildlife Service; and $7.3 million, or about 6 percent, for U.S. Department of the Interior appraisal services.

[14]    Pub. L. 105-263, 112 Stat. 2343 (1998), as amended. Other acts include the Lincoln County Land Act of 2000, Pub. L. 106-298, as amended by Pub. L. 108-424 (2004); and the White Pine County Conservation, Recreation, and Development Act of 2006, Title III, Pub. L. 109-432.

[15]   SNPLMA also authorizes the expenditure of funds on additional categories, such as certain capital improvements; development of a multispecies habitat plan in Clark County, Nevada; and development of parks, trails, and natural areas in Clark County, Nevada.

[16]   See Pub. L. 136, August 31, 1951 (65 Stat. 248, 252).

[17]   Pub. L. 106-298: Lincoln County Land Act Of 2000, as amended by Pub. L. 108-424 (2004) and Pub. L. 109-432, Title III, White Pine County Conservation, Recreation, and Development Act of 2006.

[18]   In 2000, BLM estimated that there were more than 3.3 million acres potentially available for disposal. (BLM, "Questions and Answers: Federal Land Transaction Facilitation Act, Title II, of The Valles Caldera Preservation Act (Baca Ranch, NM)" (April 4, 2003, [http://www.blm.gov/nhp/news/releases/pages/2000/valles_QsAs.htm] accessed on February 28, 2007). However, the BLM FLTFA program lead said that only a small percentage of the land designated for disposal are good candidates for sale because the projected revenue from the sale exceeds the cost to conduct the sale.

[19]   The BLM FLTFA program lead reported more recently that a revised estimate of future sales revenue had been prepared for FY 2008 and beyond. The estimate projects an average of $7.5 million in annual sales revenue or a total of $82.5 million in revenue from FY 2008 through FY 2018, with the assumptions that the program will be extended, that revised program authority adds to the inventory of land available for sale and that the program is made a priority by BLM state directors.

[20]   GAO, *Executive Guide: Effectively Implementing the Government Performance and Results Act*, GAO/GGD-96-118 (Washington, D.C.: June 1996).

[21]   Sec. 205(a) of Pub. L. 106-248 (2000), codified at 43 U.S.C. § 2305.

[22]   An acquisition is considered complete when the property title is transferred from the nonfederal landowner to the federal government.

[23]   FLTFA allows up to 20 percent of revenue raised to be used for administrative activities related to land disposals. If the agencies do not need the total amount allowed for administrative expenses, they may use the remainder for acquisitions. See figure 2.

[24]   BLM's FLTFA program lead stated that the program's emphasis on inholdings naturally addresses the management efficiency criterion because the acquisition of inholdings reduces the cost and burden of managing the public land around an inholding.

[25]   The BLM FLTFA program lead said that the criteria for acquisitions under LWCF are generally broad enough to include the criteria under FLTFA.

[26]   BLM's Assistant Director for Minerals, Realty, and Resource Protection confirmed that FLTFA "provides for two separate categories of lands...that can be purchased"—inholdings and adjacent lands—and funds must be used within these parameters.

[27]   The MOU also states that the Secretaries "may mutually decide to allocate funds to a specific acquisition project, notwithstanding [these fund allocations]."

[28]   BLM reported that the first national training workshop on FLTFA for BLM and the other three agencies was held in December 2007, during which the tracking of the funding allocations was to be clarified.

# INDEX

## A

access, ix, 6, 7, 23, 45, 48
accounting, x, 55
acquisitions, viii, ix, x, 2, 3, 5, 6, 7, 8, 9,
   12, 27, 28, 31, 32, 36, 38, 43, 44, 45,
   46, 47, 51, 52, 53, 55, 56, 57, 87, 88
adjoining landowners, ix
administrative, ix, 2, 5, 27, 31, 34, 39,
   41, 87
agricultural, 83
aid, 22, 30, 32, 37, 49, 87, 88
Alaska, vii, 85
amendments, 19, 24
appendix, x, 4, 12, 20
appraisals, 6, 40, 45, 56
appropriations, viii, 8
Arapaho, 30
Arizona, 1, 2, 12, 19, 20, 28, 30, 33, 38,
   39, 46, 48, 57, 58, 60, 78, 85
assessment, 6
assumptions, 87
attention, 38, 46
auditing, x, 58
authority, viii, 6, 47, 85, 86, 87
availability, 1, 3, 12, 17, 18, 21, 44

## B

benefits, 13, 40, 41, 45
birds, 36
Bureau of Land Management (BLM) ,
   vii, viii, ix, x, xi, 1, 2, 3, 5, 6, 7, 8, 9,
   10, 11, 12, 13, 14, 15, 17, 18, 19, 20,
   21, 22, 23, 24, 27, 28, 29, 30, 31, 32,
   33, 34, 35, 36, 37, 38, 39, 40, 41, 43,
   44, 45, 46, 47, 49, 51, 55, 56, 57, 58,
   75, 77, 84, 85, 86, 87, 88
buffer, 25, 78

## C

California, x, 1, 2, 12, 19, 20, 28, 29, 30,
   33, 37, 38, 39, 48, 49, 56, 57, 58, 60,
   78, 85
candidates, 87
capital, 87
coding, 57
Colorado, x, 1, 12, 18, 19, 20, 28, 29, 36,
   38, 39, 46, 48, 55, 57, 58, 61, 79, 85
commercial, 14
Committee on Appropriations, vii
community (ies), viii, 21, 25, 36, 82, 83
compatibility, 6
compensation, 40, 41

complexity, 2, 3, 18, 21, 43, 44
compliance, 52
Congress, viii, 3, 7, 10, 51, 52, 53, 86
Congressional Research Service, 7
conservation, 9, 46, 85, 86
consolidation, viii, 23
contracts, 40
control, 44, 56
coordination, 10
corporations, 6
costs, 34, 41, 48
covering, 1
critical habitat, 31, 36
cultural, vii, 6, 18, 22, 23, 32, 40, 48, 56, 86

**D**

decision making, 23
decisions, 49
demand, 1, 19
Department of Agriculture, vii, 4, 55, 85
Department of the Interior, vii, 3, 10, 55, 85, 86
desert, 32
Dicks, vii
discretionary, 47
draft, 3
duties, 5

**E**

economic development, 21, 82, 83
education, 8
electronic, 56
endangered, 22, 23, 31, 32
energy, ix, 1, 17, 18, 46
environmental, 6, 18, 21, 40
equalization payments, viii, 11, 12
equities, 6
evidence, x, 57, 58
expenditures, x, 27, 39, 40, 41, 55, 56
experts, 18

**F**

failure, 51
February, vii, x, 13, 58, 85, 87
federal government, 6, 45, 46, 87
Federal Land Transaction Facilitation
    Act (FLTFA) , viii, ix, x, xi, 1, 2, 3, 5,
    6, 7, 8, 9, 10, 11, 12, 13, 14, 15, 17,
    18, 19, 20, 21, 22, 23, 24, 27, 28, 29,
    30, 31, 32, 33, 35, 36, 37, 38, 39, 40,
    41, 43, 44, 45, 46, 47, 48, 49, 51, 52,
    53, 55, 56, 57, 58, 59, 77, 85, 86, 87,
    88
Federal Register, 10, 23, 37, 38, 56
fee (s), 30, 86
financial system, 56
Fish and Wildlife Service, vii, ix, 3, 5, 7,
    9, 28, 29, 30, 36, 49, 55, 56, 85, 86
flexibility, 3, 49, 52
Forest Service, vii, ix, 3, 5, 7, 9, 28, 30,
    32, 34, 38, 45, 49, 55, 56, 85, 86
forests, viii, 9
fowl, 36
fuel, 86
funding, viii, 3, 7, 8, 28, 29, 30, 34, 37,
    38, 44, 46, 47, 48, 49, 86, 88
funds, viii, ix, x, 2, 3, 5, 7, 8, 12, 13, 17,
    27, 31, 38, 39, 43, 46, 49, 51, 52, 53,
    55, 87, 88

**G**

gas, vii, 22, 86
gauge, 23
Georgia, 62
goals, 2, 3, 17, 20, 22, 23, 51, 53, 55
government, x, 6, 7, 19, 25, 44, 45, 46,
    58, 87
Government Accountability Office
    (GAO), 1, 7, 9, 12, 13, 14, 15, 20, 21,
    24, 30, 31, 34, 36, 39, 40, 58, 75, 84,
    87

Government Performance and Results
   Act, 23, 87
grazing, 6
groups, 5
growth, 24
guidance, x, 49, 55, 56

## H

habitat, 23, 31, 32, 36, 87
hazardous materials, 23, 46
House, vii

## I

Idaho, 1, 2, 12, 18, 19, 20, 23, 28, 29, 30,
   36, 38, 39, 45, 48, 57, 58, 62, 80, 85
identification, 47
implementation, ix, x, 2, 10, 17, 28, 37,
   38, 48, 51, 52, 57
incentive, 19, 41
inefficiency, 78
internal controls, 56
interpretation, 55
interview (s), x, 22, 57

## J

January, 2, 27, 28, 29, 30, 38
judgment, 22
jurisdiction (s), 46, 86
justification, 22

## L

labor, 56
Lafayette, 70
land, vii, viii, ix, x, 1, 2, 3, 5, 6, 7, 8, 9,
   10, 11, 12, 13, 14, 17, 18, 19, 21, 22,
   23, 24, 27, 28, 30, 31, 32, 33, 34, 36,
   37, 38, 39, 41, 43, 44, 45, 46, 47, 49,
   51, 52, 53, 56, 57, 58, 79, 85, 86, 87

land acquisition, ix, x, 2, 3, 7, 8, 9, 10,
   12, 27, 28, 32, 36, 37, 38, 43, 44, 45,
   46, 47, 49, 51, 52, 53, 56, 58, 86
Land and Water Conservation Fund
   (LWCF), xi, 7, 8, 28, 30, 34, 44, 47,
   52, 86, 88
land disposal, ix, 5, 22, 87
land use, viii, ix, 2, 5, 6, 10, 17, 18, 19,
   23, 24, 52, 53, 85
landfill, 78
language, 47
law (s), 3, 6, 7, 8, 18, 44, 46, 55
lead, 5, 30, 31, 32, 37, 41, 49, 87, 88
Lincoln, 19, 86, 87
Livestock, 63
local government, 6, 19, 25, 86
location, 35

## M

management, vii, viii, ix, x, 2, 10, 17, 22,
   23, 27, 47, 48, 51, 55, 56, 80, 85, 87
market, 6, 13, 14, 19, 20, 60, 61, 62, 63,
   64, 65, 66, 67, 68, 69, 70, 71, 72, 73,
   74, 75, 78, 79, 80, 81, 82, 83, 84
market value, 6, 13, 14, 20, 78, 79, 80,
   81, 82, 83, 84
memorandum of understanding (MOU) ,
   ix, xi, 2, 3, 10, 19, 27, 28, 37, 41, 43,
   47, 49, 52, 57, 88
Mexico, x, 1, 2, 12, 20, 21, 24, 27, 28,
   29, 30, 33, 34, 36, 38, 39, 46, 48, 56,
   57, 58, 69, 82, 85, 86
migratory birds, 36
mining, 23
missions, 23
misunderstanding, 49
money, 38
Montana, 1, 2, 12, 20, 28, 29, 36, 38, 39,
   48, 57, 58, 63, 85

## N

naming, 6
national, viii, ix, 2, 3, 9, 10, 20, 23, 34, 37, 43, 45, 47, 49, 57, 85, 88
National Park Service, vii, 85
national parks, viii, 9
National Wildlife Refuge, 29, 30
natural, vii, 6, 18, 48, 86, 87
negotiating, 7, 32, 34, 44
Nevada, vii, x, xi, 1, 2, 3, 8, 11, 12, 13, 14, 19, 20, 22, 27, 32, 33, 34, 35, 36, 37, 38, 39, 43, 45, 46, 47, 48, 56, 57, 58, 63, 64, 69, 81, 85, 87
New Mexico, x, 1, 2, 12, 20, 21, 24, 27, 28, 29, 30, 33, 34, 36, 38, 39, 46, 48, 56, 57, 58, 69, 82, 85, 86
Nominations, 32, 33

## O

oil, vii, 22, 86
opposition, 2, 3, 18, 19, 44, 46
Oregon, x, 1, 2, 12, 14, 20, 28, 29, 36, 38, 39, 46, 48, 49, 56, 57, 58, 70, 85
organizations, 6, 23, 58
oversight, 55
ownership, viii, 6

## P

Park Service, ix, 3, 5, 7, 9, 28, 30, 36, 38, 49, 55, 56, 86
payroll, 40
performance, 3, 22, 23, 43, 51
permit, 6
pipelines, 18
planning, 2, 18, 19, 20, 23
political, 19
population, 1, 11
preference, ix
preparation, 6, 18
prices, 14, 45

priorities, 1, 3, 7, 9, 10, 37, 38, 43, 46, 47, 51
private, vii, viii
probability, 56
procedures, 3, 10, 12, 37, 43, 56, 57
program, ix, x, 1, 2, 3, 5, 10, 11, 12, 17, 18, 22, 23, 28, 30, 31, 32, 37, 39, 41, 45, 47, 49, 55, 57, 58, 87, 88
promote, viii
property, 18, 30, 32, 36, 45, 86, 87
protection, 86
protocol, 57
public, vii, viii, ix, 2, 3, 6, 10, 18, 19, 21, 23, 36, 44, 46, 48, 51, 79, 82, 83, 86, 87
public interest, 82, 83
public notice, 10
public sector, 23

## R

range, 58, 80
real estate, 39, 45
recreation, 7, 86
recreational, 32, 36, 48, 86
regional, 58
regulations, 3, 6, 7, 18, 44, 46, 55
reliability, 55, 56, 57
reproduction, 40
resale, 6
residential, 14, 21, 79, 80
resolution, 78
resource management, viii, 80, 85
resources, vii, viii, 5, 6, 7, 18, 23, 24, 32, 36, 38, 51, 52
responsibilities, ix, 10
restoration, 10
retention, 10
revenue, viii, ix, x, 1, 2, 3, 5, 8, 11, 12, 13, 14, 17, 18, 19, 20, 22, 23, 24, 27, 31, 37, 38, 39, 41, 43, 46, 51, 52, 53, 55, 57, 58, 85, 86, 87
rights-of-way, ix, 1, 17, 18
riparian, 31, 32

roadmap, 17

## S

sales, viii, ix, x, 1, 2, 3, 5, 6, 8, 10, 11, 12, 13, 14, 15, 17, 18, 19, 20, 21, 22, 23, 25, 27, 31, 39, 41, 51, 53, 55, 56, 57, 86, 87
sample, x, 56, 57
scientific, 48, 86
Secretary of Agriculture, 6
security, viii
services, 40, 57, 86
sewage, 83
sharing, 5
shorebirds, 36
sites, x, 56
software, 57
South Dakota, 63
species, 22, 23, 32, 36, 48
specificity, 6
staffing, 18, 22
standards, x, 58
state office, ix, x, 1, 2, 5, 11, 12, 13, 14, 15, 17, 20, 21, 22, 24, 32, 34, 38, 44, 49, 56, 57, 75, 84
statistics, 57
statutory, 47
surplus, 86
systems, 47, 52, 56

## T

targets, 22, 49
tax base, 46
third party, 44
threatened, 22, 23, 31, 32
time, ix, 2, 3, 7, 18, 19, 21, 27, 39, 43, 44, 51, 52
Title III, 86, 87
total revenue, 20
tracking, 88

training, 18, 88
transactions, viii, 11, 13, 56, 85
transfer, 80
transmission, 18
transportation, 8, 40
Treasury, viii, ix, 10, 23, 56, 85
trust fund, 86
turnover, 18

## U

U.S. Department of Agriculture, vii, 55, 85
U.S. Treasury, viii, ix, 10, 23, 56, 85
United States, vii
urban areas, 57
users, ix, 6
Utah, 1, 12, 20, 21, 36, 37, 38, 39, 44, 48, 57, 58, 75, 83, 85

## V

values, 14, 20, 45, 46
Virginia, 74

## W

Washington, vii, 1, 12, 20, 38, 39, 48, 57, 85, 87
waste disposal, 80
water resources, 36
water rights, 36
web-based, 57
wetlands, 23
wildlife, viii, 31, 36
workforce, ix
workload, 18
writing, 55
Wyoming, 1, 2, 12, 20, 22, 28, 29, 30, 34, 38, 39, 44, 48, 57, 58, 75, 84, 85